COLLINS
WORLD
POCKET
DataFile

D0366092

Bartholomew
A Division of HarperCollins Publishers
Duncan Street, Edinburgh EH9 1TA

© Bartholomew 1993

First published by Bartholomew 1993

ISBN 0 00 448070 8

Produced by HarperCollins, Hong Kong

Details included in this book are subject to change without notice.
Whilst every effort is made to keep information up to date HarperCollins
Publishers will not be responsible for any loss, damage or
inconvenience caused by inaccuracies in this book. The publishers are
always pleased to acknowledge any corrections brought to their notice,
and record their appreciation of the valuable services rendered in the
past by map users in assisting to maintain the accuracy of their
publications.

F/E 6904

COLLINS
WORLD
POCKET
DataFile

HarperCollins*Publishers*

WORLD POCKET DATAFILE

On pages 28 to 236 of the book, countries are arranged in alphabetical order, except colonies or dependencies which appear after the parent country.

The statistics quoted are from latest available sources, including UN data. The languages and religions given are the official or dominant ones of the country. In many countries more than one language is spoken and in most countries several different religions are followed.

The abbreviations used for international organisation membership are explained below.

AL	Arab League
ALADI	Latin American Integration Association
ASEAN	Association of South-East Asian Nations
C	Commonwealth
CACM	Central American Common Market
CARICOM	Caribbean Community
CP	Colombo Plan
EC	European Community
ECOWAS	Economic Community of West African States
EFTA	European Free Trade Association
NATO	North Atlantic Treaty Organisation
OAS	Organisation of American States
OAU	Organisation of African Unity
OECD	Organisation for Economic Co-operation and Development
OPEC	Organisation of Petroleum Exporting Countries
PTA	Eastern and Southern African Preferential Trade Area
SADC	Southern African Development Community
SPF	South Pacific Forum
UDEAC	Customs and Economic Union of Central Africa
UN	United Nations

CONTENTS

WORLD

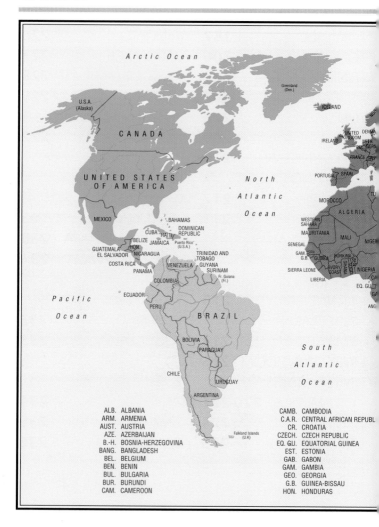

Arctic Ocean

Greenland (Den.)

ICELAND

U.S.A. (Alaska)

CANADA

UNITED KINGDOM
IRELAND
DENMARK
NETH.
BEL. GERM.
FRANCE SWI.

North
Atlantic
Ocean

UNITED STATES
OF AMERICA

PORTUGAL SPAIN

MEXICO

BAHAMAS
DOMINICAN
REPUBLIC
CUBA HAITI
BELIZE JAMAICA Puerto Rico (U.S.A.)
GUATEMALA HON.
EL SALVADOR NICARAGUA
COSTA RICA
PANAMA

MOROCCO

ALGERIA

WESTERN
SAHARA
MAURITANIA
MALI
NIGER
SENEGAL
GAM.
G.B. GUINEA
SIERRA LEONE
LIBERIA
IVORY
COAST
BURKINA
GHANA
NIGERIA
EQ. GU.
GA
ANG.

TRINIDAD AND
TOBAGO
VENEZUELA GUYANA
SURINAM
COLOMBIA Fr. Guiana (Fr.)

Pacific
Ocean

ECUADOR

PERU

BRAZIL

South
Atlantic
Ocean

BOLIVIA
PARAGUAY

CHILE

URUGUAY

ARGENTINA

Falkland Islands (U.K)

ALB.	ALBANIA	CAMB.	CAMBODIA
ARM.	ARMENIA	C.A.R.	CENTRAL AFRICAN REPUBL
AUST.	AUSTRIA	CR.	CROATIA
AZE.	AZERBAIJAN	CZECH.	CZECH REPUBLIC
B.-H.	BOSNIA-HERZEGOVINA	EQ. GU.	EQUATORIAL GUINEA
BANG.	BANGLADESH	EST.	ESTONIA
BEL.	BELGIUM	GAB.	GABON
BEN.	BENIN	GAM.	GAMBIA
BUL.	BULGARIA	GEO.	GEORGIA
BUR.	BURUNDI	G.B.	GUINEA-BISSAU
CAM.	CAMEROON	HON.	HONDURAS

12

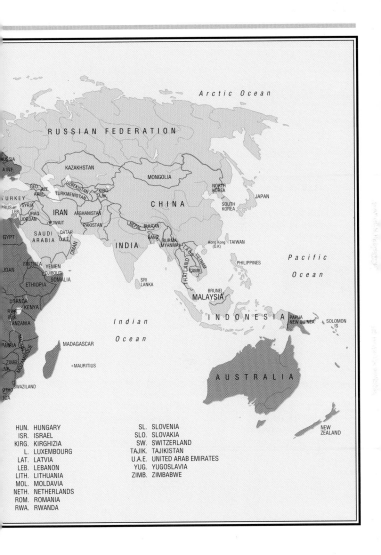

HUN. HUNGARY
ISR. ISRAEL
KIRG. KIRGHIZIA
L. LUXEMBOURG
LAT. LATVIA
LEB. LEBANON
LITH. LITHUANIA
MOL. MOLDAVIA
NETH. NETHERLANDS
ROM. ROMANIA
RWA. RWANDA

SL. SLOVENIA
SLO. SLOVAKIA
SW. SWITZERLAND
TAJIK. TAJIKISTAN
U.A.E. UNITED ARAB EMIRATES
YUG. YUGOSLAVIA
ZIMB. ZIMBABWE

EUROPE

EUROPE : POPULATION

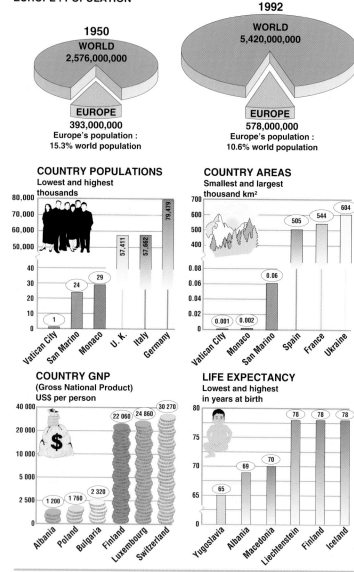

1950

WORLD
2,576,000,000

EUROPE
393,000,000
Europe's population :
15.3% world population

1992

WORLD
5,420,000,000

EUROPE
578,000,000
Europe's population :
10.6% world population

COUNTRY POPULATIONS
Lowest and highest
thousands

Vatican City	San Marino	Monaco	U. K.	Italy	Germany
1	24	29	57,411	57,662	79,479

COUNTRY AREAS
Smallest and largest
thousand km²

Vatican City	Monaco	San Marino	Spain	France	Ukraine
0.001	0.002	0.06	505	544	604

COUNTRY GNP
(Gross National Product)
US$ per person

Albania	Poland	Bulgaria	Finland	Luxembourg	Switzerland
1 200	1 760	2 320	22 060	24 860	30 270

LIFE EXPECTANCY
Lowest and highest
in years at birth

Yugoslavia	Albania	Macedonia	Liechtenstein	Finland	Iceland
65	69	70	78	78	78

Greenland
(Den.)

Svalbard

A r c t i c   O c e a n

ICELAND

Atlantic
North Sea

UNITED
IRELAND
KINGDOM

Ocean

FRANCE

PORTUGAL
SPAIN

5

Mediterranean   Sea

MALTA

0   400 km

NORWAY
SWEDEN
FINLAND

ESTONIA
LATVIA
LITHUANIA
10

DENMARK
1
BELORUSSIA

POLAND
GERMANY
2
3   11   UKRAINE
SLOVAKIA
AUSTRIA   18
4   6   HUNGARY
12 13   ROMANIA   Black Sea
7   8   14
I   15   BULGARIA
T   16   TURKEY
A
L   17
Y   9   GREECE
CYPRUS

1	NETHERLANDS	7	MONACO	13	CROATIA
2	BELGIUM	8	SAN MARINO	14	BOSNIA-HERZEGOVINA
3	LUXEMBOURG	9	VATICAN CITY	15	YUGOSLAVIA
4	SWITZERLAND	10	RUSSIAN FEDERATION	16	MACEDONIA
5	ANDORRA	11	CZECH REPUBLIC	17	ALBANIA
6	LIECHTENSTEIN	12	SLOVENIA	18	MOLDAVIA

EUROPE is the smallest of the continents in area, and, because of low birth rates, has a decreasing proportion of the world's population. Only USA and Japan have higher GNP's than those of the richest countries in Europe.

ASIA

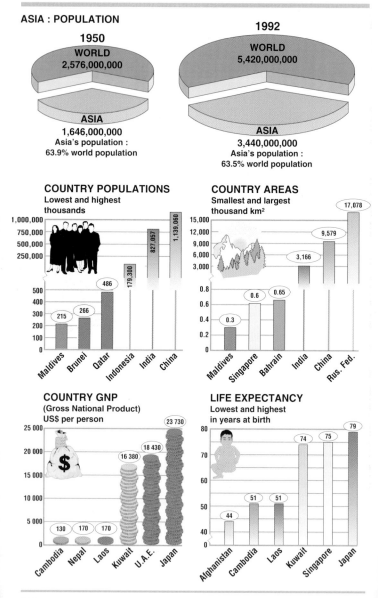

ASIA : POPULATION

1950
WORLD
2,576,000,000

ASIA
1,646,000,000
Asia's population :
63.9% world population

1992
WORLD
5,420,000,000

ASIA
3,440,000,000
Asia's population :
63.5% world population

COUNTRY POPULATIONS
Lowest and highest
thousands

1,000,000
750,000
500,000
250,000

500 — 486
400
300 — 266
200 — 215
100
0

Maldives	Brunei	Qatar	Indonesia	India	China
215	266	486	179,300	827,057	1,139,060

COUNTRY AREAS
Smallest and largest
thousand km²

15,000 — 17,078
12,000
9,000 — 9,579
6,000
3,000 — 3,166

0.8
0.6 — 0.6   0.65
0.4
0.2 — 0.3
0

Maldives	Singapore	Bahrain	India	China	Rus. Fed.
0.3	0.6	0.65	3,166	9,579	17,078

COUNTRY GNP
(Gross National Product)
US$ per person

25 000 — 23 730
20 000 — 18 430
15 000 — 16 380
10 000
5 000

Cambodia	Nepal	Laos	Kuwait	U.A.E.	Japan
130	170	170	16 380	18 430	23 730

LIFE EXPECTANCY
Lowest and highest
in years at birth

80 — 79
74   75
70
60
51   51
50 — 44
40

Afghanistan	Cambodia	Laos	Kuwait	Singapore	Japan
44	51	51	74	75	79

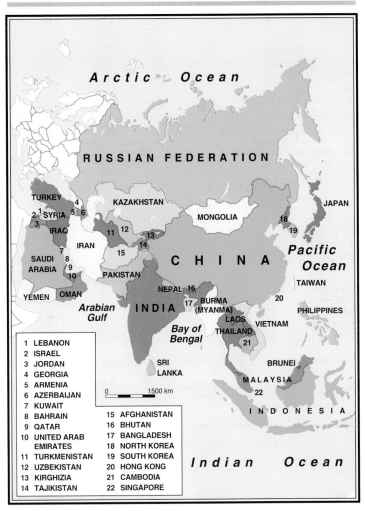

1 LEBANON
2 ISRAEL
3 JORDAN
4 GEORGIA
5 ARMENIA
6 AZERBAIJAN
7 KUWAIT
8 BAHRAIN
9 QATAR
10 UNITED ARAB EMIRATES
11 TURKMENISTAN
12 UZBEKISTAN
13 KIRGHIZIA
14 TAJIKISTAN
15 AFGHANISTAN
16 BHUTAN
17 BANGLADESH
18 NORTH KOREA
19 SOUTH KOREA
20 HONG KONG
21 CAMBODIA
22 SINGAPORE

ASIA is the largest of the continents, in area and population. China alone has one fifth of the world's population. Many countries are poor, but some are industrialising along the 'Pacific Rim', and the Middle East has much of the world's oil.

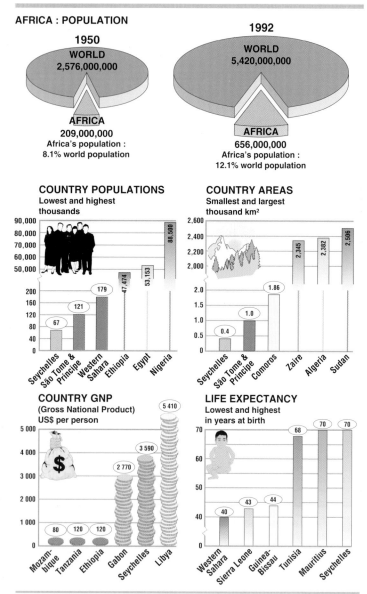

AFRICA : POPULATION

1950
WORLD
2,576,000,000

AFRICA
209,000,000
Africa's population :
8.1% world population

1992
WORLD
5,420,000,000

AFRICA
656,000,000
Africa's population :
12.1% world population

COUNTRY POPULATIONS
Lowest and highest
thousands

- Seychelles: 67
- São Tome & Principe: 121
- Western Sahara: 179
- Ethiopia: 47,474
- Egypt: 53,153
- Nigeria: 88,500

COUNTRY AREAS
Smallest and largest
thousand km²

- Seychelles: 0.4
- São Tome & Principe: 1.0
- Comoros: 1.86
- Zaire: 2,345
- Algeria: 2,382
- Sudan: 2,506

COUNTRY GNP
(Gross National Product)
US$ per person

- Mozambique: 80
- Tanzania: 120
- Ethiopia: 120
- Gabon: 2 770
- Seychelles: 3 590
- Libya: 5 410

LIFE EXPECTANCY
Lowest and highest
in years at birth

- Western Sahara: 40
- Sierra Leone: 43
- Guinea-Bissau: 44
- Tunisia: 68
- Mauritius: 70
- Seychelles: 70

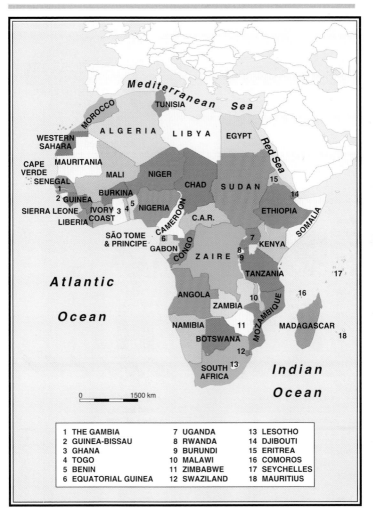

1 THE GAMBIA	7 UGANDA	13 LESOTHO
2 GUINEA-BISSAU	8 RWANDA	14 DJIBOUTI
3 GHANA	9 BURUNDI	15 ERITREA
4 TOGO	10 MALAWI	16 COMOROS
5 BENIN	11 ZIMBABWE	17 SEYCHELLES
6 EQUATORIAL GUINEA	12 SWAZILAND	18 MAURITIUS

AFRICA is the poorest continent. The population is increasing rapidly in every country, despite widespread civil wars and outbreaks of famine. These problems are hindering development in many countries.

NORTH & CENTRAL AMERICA : POPULATION

1950

WORLD
2,576,000,000

N & C AMERICA
216,000,000
North & Central America's
population : 8.4% world population

1992

WORLD
5,420,000,000

N & C AMERICA
432,000,000
North & Central America's
population : 8.0% world population

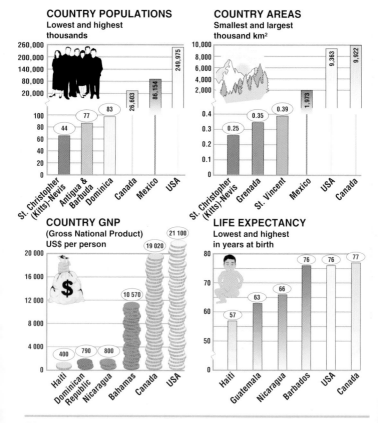

COUNTRY POPULATIONS
Lowest and highest
thousands

St. Christopher (Kitts)-Nevis	44
Antigua & Barbuda	77
Dominica	83
Canada	26,603
Mexico	86,154
USA	249,975

COUNTRY AREAS
Smallest and largest
thousand km²

St. Christopher (Kitts)-Nevis	0.25
Grenada	0.35
St. Vincent	0.39
Mexico	1,973
USA	9,363
Canada	9,922

COUNTRY GNP
(Gross National Product)
US$ per person

Haiti	400
Dominican Republic	790
Nicaragua	800
Bahamas	10 570
Canada	19 020
USA	21 100

LIFE EXPECTANCY
Lowest and highest
in years at birth

Haiti	57
Guatemala	63
Nicaragua	66
Barbados	76
USA	76
Canada	77

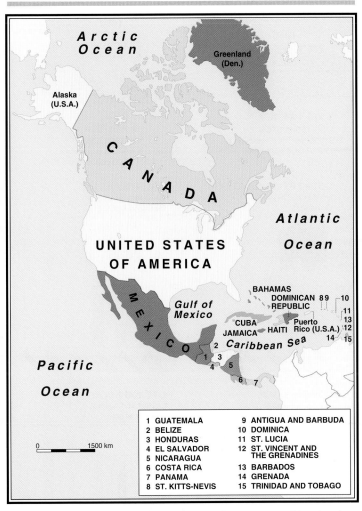

1 GUATEMALA	9 ANTIGUA AND BARBUDA
2 BELIZE	10 DOMINICA
3 HONDURAS	11 ST. LUCIA
4 EL SALVADOR	12 ST. VINCENT AND THE GRENADINES
5 NICARAGUA	
6 COSTA RICA	13 BARBADOS
7 PANAMA	14 GRENADA
8 ST. KITTS-NEVIS	15 TRINIDAD AND TOBAGO

NORTH AND CENTRAL AMERICA have been centres of immigration for 400 years. Canada and the U.S.A. have most resources and are the most developed. Central American countries are smaller and poorer .

SOUTH AMERICA

SOUTH AMERICA : POPULATION

1950

WORLD
2,576,000,000

S. AMERICA
110,000,000
South America's population :
4.3% world population

1992

WORLD
5,420,000,000

S. AMERICA
302,000,000
South America's population :
5.5% world population

COUNTRY POPULATIONS
Lowest and highest
thousands

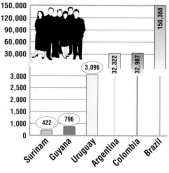

Surinam	422
Guyana	796
Uruguay	3,096
Argentina	32,322
Colombia	32,987
Brazil	150,368

COUNTRY AREAS
Smallest and largest
thousand km²

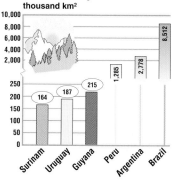

Surinam	164
Uruguay	187
Guyana	215
Peru	1,285
Argentina	2,778
Brazil	8,512

COUNTRY GNP
(Gross National Product)
US$ per person

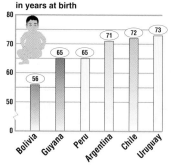

Guyana	340
Bolivia	600
Paraguay	1030
Brazil	2 550
Uruguay	2 620
Surinam	3 020

LIFE EXPECTANCY
Lowest and highest
in years at birth

Bolivia	56
Guyana	65
Peru	65
Argentina	71
Chile	72
Uruguay	73

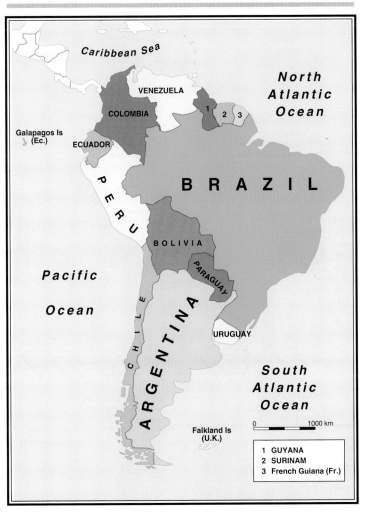

S **outh America** is still a relatively underexploited continent. Birth rates are high. Despite being resource rich the countries of South America have not yet reached their full economic potential.

AUSTRALASIA

AUSTRALASIA : POPULATION

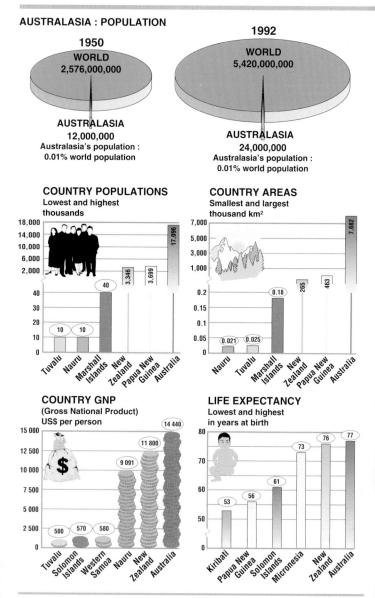

1950
WORLD
2,576,000,000

AUSTRALASIA
12,000,000
Australasia's population :
0.01% world population

1992
WORLD
5,420,000,000

AUSTRALASIA
24,000,000
Australasia's population :
0.01% world population

COUNTRY POPULATIONS
Lowest and highest
thousands

- Tuvalu: 10
- Nauru: 10
- Marshall Islands: 40
- New Zealand: 3,346
- Papua New Guinea: 3,699
- Australia: 17,096

COUNTRY AREAS
Smallest and largest
thousand km²

- Nauru: 0.021
- Tuvalu: 0.025
- Marshall Islands: 0.18
- New Zealand: 265
- Papua New Guinea: 463
- Australia: 7,682

COUNTRY GNP
(Gross National Product)
US$ per person

- Tuvalu: 500
- Solomon Islands: 570
- Western Samoa: 580
- Nauru: 9 091
- New Zealand: 11 800
- Australia: 14 440

LIFE EXPECTANCY
Lowest and highest
in years at birth

- Kiribati: 53
- Papua New Guinea: 56
- Solomon Islands: 61
- Micronesia: 73
- New Zealand: 76
- Australia: 77

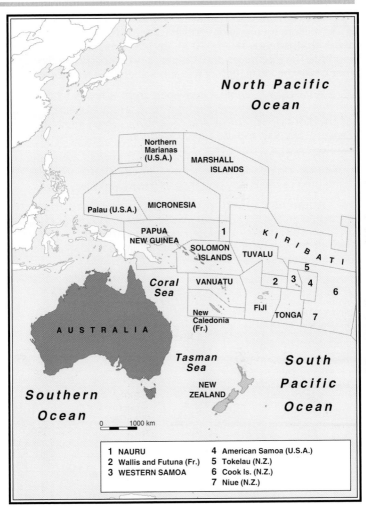

North Pacific Ocean

Northern Marianas (U.S.A.)

MARSHALL ISLANDS

Palau (U.S.A.)

MICRONESIA

PAPUA NEW GUINEA

1

K I R I B A T I

SOLOMON ISLANDS

TUVALU

5

VANUATU

2   3   4

6

New Caledonia (Fr.)

FIJI

TONGA   7

AUSTRALIA

Coral Sea

Tasman Sea

NEW ZEALAND

South Pacific Ocean

Southern Ocean

0   1000 km

1	NAURU	4	American Samoa (U.S.A.)
2	Wallis and Futuna (Fr.)	5	Tokelau (N.Z.)
3	WESTERN SAMOA	6	Cook Is. (N.Z.)
		7	Niue (N.Z.)

Australasia's share of the world's population is negligible. There is a marked contrast between the rich, developed countries of Australia and New Zealand, and the poorer island groups in the Pacific Ocean

ANTARCTICA

ANTARCTICA is an isolated island continent about 15.5 million square kilometres in area. It is covered by an ice sheet that extends nearly 1000 km into the sea in the winter months. The intense cold in Antarctica is one of the mechanisms that drives the circulation of the atmosphere and the oceans.

Antarctica is uninhabited apart from a few scientific and research workers. Several countries have territorial claims to parts of the continent. These are Argentina, Chile, France, New Zealand, Norway and the United Kingdom.

In 1959 the Antarctic Treaty was signed whereby all territorial claims are held in abeyance. There are now 39 signatories to the Treaty and they all meet bi-annually.

Antarctica is rich in natural resources and there are pressures on the commercial exploitation of the minerals in the area. In 1991 it was proposed that there should be a 50 year ban on mining, and this would be an important step towards creating a world park.

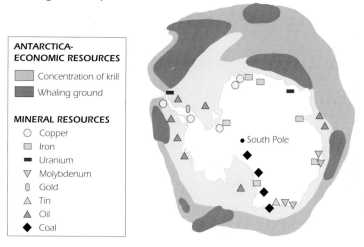

ANTARCTICA-
ECONOMIC RESOURCES

▢ Concentration of krill

▨ Whaling ground

MINERAL RESOURCES

○ Copper
▢ Iron
▬ Uranium
▽ Molybdenum
◖ Gold
△ Tin
▲ Oil
◆ Coal

• South Pole

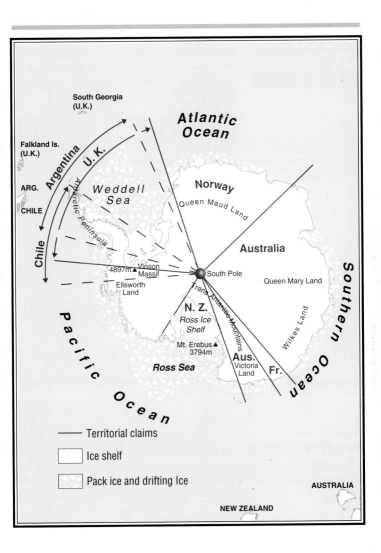

South Georgia
(U.K.)

Falkland Is.
(U.K.)

**Atlantic
Ocean**

ARG.

CHILE

Argentina

U. K.

Chile

Antarctic Peninsula

*Weddell
Sea*

Norway

Queen Maud Land

Australia

Queen Mary Land

Wilkes Land

Southern Ocean

4897m ▲ Vinson
Massif

Ellsworth
Land

South Pole

Trans Antarctic Mountains

N. Z.

*Ross Ice
Shelf*

Mt. Erebus ▲
3794m

Ross Sea

Aus.
Victoria
Land

Fr.

Pacific Ocean

—— Territorial claims

Ice shelf

Pack ice and drifting Ice

AUSTRALIA

NEW ZEALAND

AFGHANISTAN

STATUS	**Republic**
AREA	**652,225 sq km**
CAPITAL	**Kabul**
MAIN CITIES	**Kandahar, Herat**
POPULATION	**16,121,000**
Density	**25 people per sq km**
Life Expectancy	**44 years**
Infant Mortality	**162 per 1000**
LANGUAGES	**Pushtu, Dari**

RELIGIONS	**Sunni Muslim**
CURRENCY	**Afghani**
GNP	**250 US $ per person**
MAIN PRODUCTS	**Dried fruit and nuts, Carpets and rugs, Wool and hides**
NATIONAL DAY	**April 27**
ORGANISATIONS	**UN, CP**

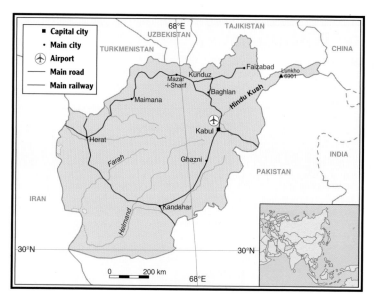

THIS **MOUNTAINOUS** landlocked country has a climate of extremes and variable rainfall. In the south-west are the lowland areas while mountains lie in the north. One of the poorest countries in the world with very little land suitable for agriculture. Main crops are wheat, fruit and vegetables. Sheep and goats are the principal livestock. Mineral resources are rich but underdeveloped with natural gas, coal and iron ore deposits predominating. The main industrial area is centred on Kabul.

STATUS **Republic**	RELIGIONS **Muslim**
AREA **28,750 sq km**	CURRENCY **Lek**
CAPITAL **Tirana**	GNP **1,200 US $ per person**
POPULATION **3,250,000**	MAIN PRODUCTS **Crude minerals, Food**
Density **113 people per sq km**	
Life Expectancy **73 years**	NATIONAL DAY **January 11**
Infant Mortality **32 per 1000**	ORGANISATIONS **UN**
LANGUAGES **Albanian**	

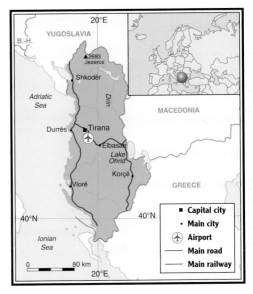

SITUATED on the eastern seaboard of the Adriatic Sea. With the exception of a coastal strip, most of the territory is mountainous and largely unfit for cultivation. Possesses considerable mineral resources, notably chrome, copper, iron ores and nickel, with rich deposits of coal, oil and natural gas. After the fall of communism in 1990, production fell massively, especially in agriculture. Acute food shortages and economic backwardness have generated a desire for emigration among the younger members of the fast-growing population.

ALGERIA

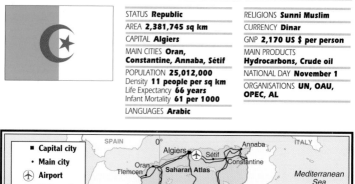

STATUS **Republic**	RELIGIONS **Sunni Muslim**
AREA **2,381,745 sq km**	CURRENCY **Dinar**
CAPITAL **Algiers**	GNP **2,170 US $ per person**
MAIN CITIES **Oran, Constantine, Annaba, Sétif**	MAIN PRODUCTS **Hydrocarbons, Crude oil**
POPULATION **25,012,000** Density **11 people per sq km** Life Expectancy **66 years** Infant Mortality **61 per 1000**	NATIONAL DAY **November 1** ORGANISATIONS **UN, OAU, OPEC, AL**
LANGUAGES **Arabic**	

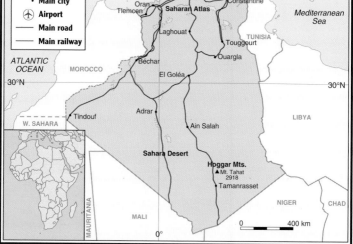

FORMER DEPARTMENT of France and a French colony; independent 1962. Stretches from the Sahara Desert in the south to coastal mountains in the north. Arable land occupies small areas of the northern valleys and coastal strip, with wheat, barley and vines the leading crops. Sheep, goats and cattle are the most important livestock. Although declining, oil from the desert still dominates the economy. Tourism is a growth industry and earns important foreign exchange.

AMERICAN SAMOA see page 223

STATUS **Principality**	RELIGIONS **Roman Catholic**
AREA **465 sq km**	CURRENCY **French Franc, Spanish Peseta**
CAPITAL **Andorra-la-Vella**	
POPULATION **52,000**	GNP **16,600 US $ per person**
Density **112 people per sq km**	MAIN PRODUCTS **Clothing, Food**
Life Expectancy **78 years**	
Infant Mortality **13 per 1000**	NATIONAL DAY **September 8**
LANGUAGES **Catalan**	

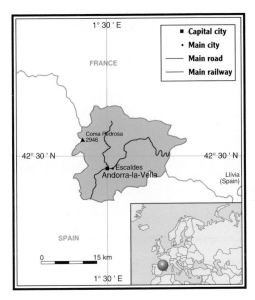

SELF-GOVERNING principality under joint sovereignty of France and Spain. This tiny alpine state lies high in the Pyrenees between France and Spain. Agriculture and tourism are the main activities. Tobacco and potatoes are the principal crops; sheep and cattle the main livestock.

ANGOLA

STATUS **Republic**	RELIGIONS **Roman Catholic**
AREA **1,246,700 sq km**	CURRENCY **Kwanza**
CAPITAL **Luanda**	GNP **620 US $ per person**
MAIN CITIES **Lubango**	MAIN PRODUCTS **Crude oil, Diamonds**
POPULATION **10,020,000** Density **8 people per sq km** Life Expectancy **46 years** Infant Mortality **127 per 1000**	NATIONAL DAY **November 11**
	ORGANISATIONS **UN, SADC**
LANGUAGES **Portuguese**	

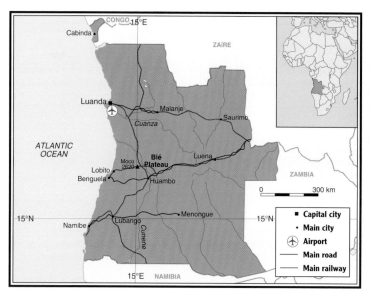

INDEPENDENT from Portugal 1975. Large country south of the equator in south-western Africa. Much of the interior is savannah plateau with variable rainfall. Most of the population is engaged in agriculture producing cassava, maize and coffee. Very rich in diamonds, iron ore, copper and manganese as well as oil. The small amount of industry is concentrated around Luanda. Most consumer products are imported.

ANGUILLA see page 217

ANTIGUA AND BARBUDA

STATUS **Constitutional Monarchy**

AREA **442 sq km**

CAPITAL **St John's**

POPULATION **77,000**
Density **174 people per sq km**
Life Expectancy **73 years**
Infant Mortality **21 per 1000**

LANGUAGES **English**

RELIGIONS **Protestant**

CURRENCY **EC Dollar**

GNP **2,800 US $ per person**

MAIN PRODUCTS **Rum, Cotton lint**

NATIONAL DAY **November 1**

ORGANISATIONS **UN, OAS, CARICOM, C**

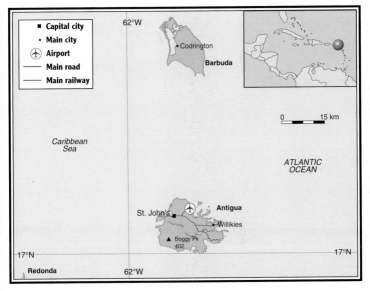

UK COLONY since 1632; independent 1984. Consists of two main islands in the Leeward group in the West Indies. Tourism is the main activity but local agriculture is being encouraged to reduce food imports. Rum production is the main manufacturing industry.

ARGENTINA

STATUS **Federal Republic**	LANGUAGES **Spanish**
AREA **2,777,815 sq km**	RELIGIONS **Roman Catholic**
CAPITAL **Buenos Aires**	CURRENCY **Austral**
MAIN CITIES **Córdoba, Rosario, Mendoza, La Plata, San Miguel de Tucumán**	GNP **2,160 US $ per person**
	MAIN PRODUCTS **Cereals, Animal feed, Machinery, Transport equipment**
POPULATION **32,322,000** Density **12 people per sq km** Life Expectancy **71 years** Infant Mortality **29 per 1000**	NATIONAL DAY **May 25**
	ORGANISATIONS **UN, OAS, ALADI**

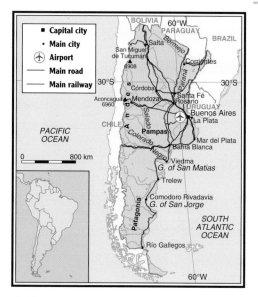

INDEPENDENT from Spain 1816. Stretches from subtropical forest in the north to the grassy plains of the pampas and the desert plateaux of Patagonia in the south. Economy was dominated by grain and beef but these declined in the 1980s owing to competition from western Europe and falling grain prices. Industry also declined in the 1980s. Expansion of oil and gas industry and growth of coal, hydro-electricity and nuclear power is providing the basis for industrial growth but inflation remains a problem.

STATUS **Republic**	RELIGIONS **Orthodox**
AREA **29,800 sq km**	CURRENCY **Rouble**
CAPITAL **Yerevan**	MAIN PRODUCTS **Minerals, Chemicals, Cotton**
MAIN CITIES **Kumayri, Karaklis**	
POPULATION **3,324,000**	ORGANISATIONS **UN**
Density **112 people per sq km**	
Life Expectancy **72 years**	
Infant Mortality **20 per 1000**	
LANGUAGES **Armenian**	

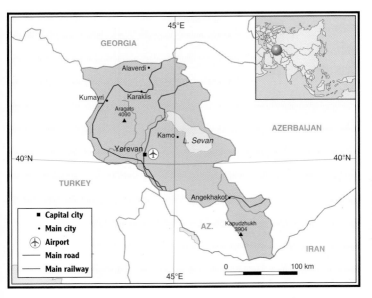

INDEPENDENT 1991. Smallest of the 15 republics of the former USSR. This rugged and landlocked country has hot, dry summers and severe winters. Arable land is limited but fertile. Extensive mountain pastures support cattle, sheep and goats. Industry is concerned mainly with machine-building, chemicals and textiles. Conflict with neighbouring Azerbaijan looks set to cast a cloud over the immediate future.

ARUBA see page 155

STATUS	**Federal State**
AREA	**7,682,300 sq km**
CAPITAL	**Canberra**
MAIN CITIES	**Sydney, Melbourne, Brisbane, Adelaide, Perth**
POPULATION	**17,086,000**
Density	**2 people per sq km**
Life Expectancy	**77 years**
Infant Mortality	**7 per 1000**
LANGUAGES	**English**
RELIGIONS	**Protestant, Roman Catholic**
CURRENCY	**Dollar**
GNP	**14,440 US $ per person**
MAIN PRODUCTS	**Metal ores, Coal, Cereals, Meat products**
NATIONAL DAY	**January 26**
ORGANISATIONS	**UN, OECD, SPF, C, CP**

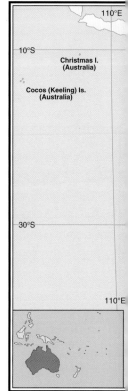

INDEPENDENT from UK 1901. Both a continent and a country; sixth largest country in the world by area. Subtropical in the north but the centre and west are desert and scrub. Most people live in the cities and towns around the coast. Agriculture still important despite growth in mineral exploitation. Vast reserves of coal, oil, natural gas, nickel, iron ore and bauxite. Trade with eastern Asia, in particular Japan, has increased, as trade with Europe has declined.

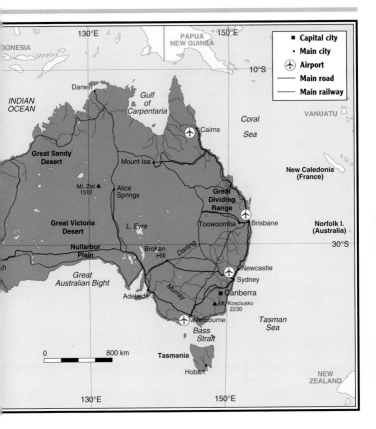

Legend
- ■ Capital city
- · Main city
- ✈ Airport
- —— Main road
- —— Main railway

INDONESIA

PAPUA NEW GUINEA

130°E · 150°E

10°S

INDIAN OCEAN

Darwin

Gulf of Carpentaria

Coral Sea

VANUATU

✈ Cairns

Great Sandy Desert

Mount Isa

New Caledonia (France)

Mt. Ziel ▲ 1510

Alice Springs

Great Dividing Range

Great Victoria Desert

L. Eyre

Toowoomba ✈ Brisbane

Norfolk I. (Australia)

Nullarbor Plain

Broken Hill

Darling

30°S

Great Australian Bight

Murray

Newcastle
✈ Sydney

Adelaide

■ Canberra
▲ Mt. Kosciusko 2230

Tasman Sea

✈ Melbourne

Bass Strait

0 — 800 km

Tasmania

Hobart

NEW ZEALAND

130°E · 150°E

AUSTRIA

STATUS **Federal Republic**	RELIGIONS **Roman Catholic**
AREA **83,855 sq km**	CURRENCY **Schilling**
CAPITAL **Vienna**	GNP **17,360 US $ per person**
MAIN CITIES **Graz, Linz, Salzburg, Innsbruck**	MAIN PRODUCTS **Machinery, Transport equipment, Chemicals**
POPULATION **7,712,000** Density **92 people per sq km** Life Expectancy **75 years** Infant Mortality **9 per 1000**	NATIONAL DAY **October 26**
	ORGANISATIONS **UN, EFTA, OECD**
LANGUAGES **German**	

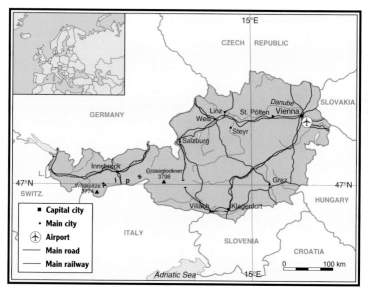

OCCUPIED by Germany 1938-45; freely elected government under occupation by Allies 1945-55; independent 1955. Mountainous Alps consist of a series of east-west ranges; climate is continental with cold winters and warm summers. In the north and north-east lie the most fertile farmlands. Half is pasture; remainder is mainly for root or fodder crops. Manufacturing and heavy industry, particularly pig-iron, steel, chemicals and vehicles, are major export earners. Tourism and forestry are also important to the economy.

STATUS **Republic**	RELIGIONS **Shi'a Muslim**
AREA **86,600 sq km**	CURRENCY **Manat**
CAPITAL **Baku**	MAIN PRODUCTS **Crude oil, Clothing, Textiles, Footwear**
MAIN CITIES **Gyandzha, Stepanakert**	ORGANISATIONS **UN**
POPULATION **7,153,000** Density **83 people per sq km** Life Expectancy **70 years** Infant Mortality **26 per 1000**	
LANGUAGES **Azerbaijani**	

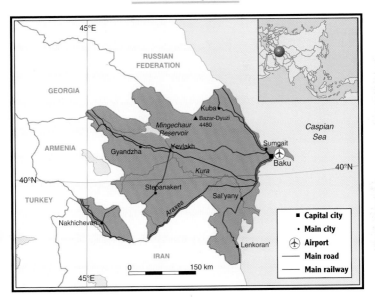

INDEPENDENT 1991. Comprises two autonomous regions separated by a strip of Armenian territory. Benefits from dry, subtropical climate with mild winters and long, hot summers. Oil extraction and refining are now supplemented by manufacturing, engineering and chemicals. Raw cotton, tobacco and grapes are leading agricultural products. Economic adaptation to post-communist market conditions is made more difficult by political uncertainty.

AZORES see page 171

BAHAMAS

STATUS **Constitutional Monarchy**

AREA **13,865 sq km**

CAPITAL **Nassau**

MAIN CITIES **Freeport**

POPULATION **253,000**
Density **18 people per sq km**
Life Expectancy **70 years**
Infant Mortality **21 per 1000**

LANGUAGES **English**

RELIGIONS **Protestant**

CURRENCY **Dollar**

GNP **10,570 US $ per person**

MAIN PRODUCTS **Chemicals, Refined oil, Hormones, Crayfish**

NATIONAL DAY **July 10**

ORGANISATIONS **UN, OAS, CARICOM, C**

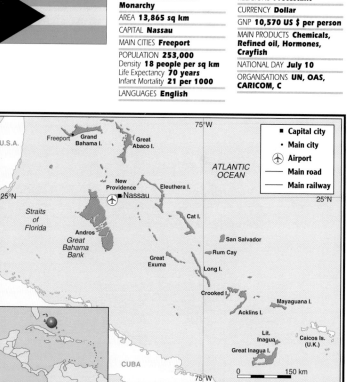

INTERNAL self-government 1964; independent from UK 1973. Comprises about 700 subtropical islands and over 2,000 corals and cays (reefs). Only 29 islands are inhabited. Most rainfall occurs in summer. Tourism is a major employer. Other important sources of income are from ship registration and off-shore finance and banking. Recent economic plans have concentrated on reducing imports by developing fishing and domestic agriculture.

STATUS **Monarchy**	RELIGIONS **Muslim**
AREA **661 sq km**	CURRENCY **Dinar**
CAPITAL **Manama**	GNP **6,360 US $ per person**
MAIN CITIES **Muharraq**	MAIN PRODUCTS **Petroleum products, Aluminium products**
POPULATION **503,000** Density **761 people per sq km** Life Expectancy **72 years** Infant Mortality **12 per 1000**	
	NATIONAL DAY **December16**
	ORGANISATIONS **UN, AL**
LANGUAGES **Arabic**	

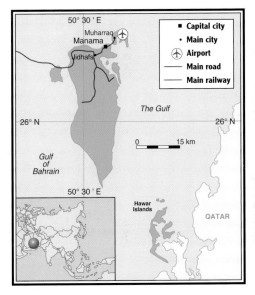

FORMER UK protected state; independent 1971. Barren island in The Gulf with minimal rainfall. First country in the Arabian Peninsula to strike oil, in 1930s. Decline in value of oil and lower production since 1980s have encouraged government to diversify the economy. Light and heavy industry and chemical plants have all been expanded, and trade and foreign investment encouraged.

BANGLADESH

STATUS **Republic**	RELIGIONS **Muslim**
AREA **144,000 sq km**	CURRENCY **Taka**
CAPITAL **Dhaka**	GNP **180 US $ per person**
MAIN CITIES **Chittagong, Khulna, Rajshaji**	MAIN PRODUCTS **Textiles, Leather, Fish, Tea, Jute**
POPULATION **115,594,000**	NATIONAL DAY **March 26**
Density **803 people per sq km**	ORGANISATIONS **UN, C, CP**
Life Expectancy **53 years**	
Infant Mortality **108 per 1000**	
LANGUAGES **Bengali**	

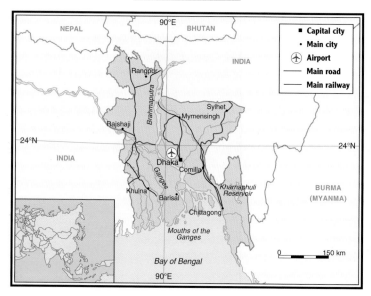

PART OF British India until 1947; independent from Pakistan 1971. One of the poorest and most densely populated countries in the world. Except for bamboo-forested hills in the south-east, country comprises the vast river systems of the Ganges and Brahmaputra, which frequently flood the flat delta plain. Agriculture is dependent on monsoon rainfall and the delta plain is very fertile, the main crops being rice and jute. No great mineral deposits exist but large reserves of natural gas are being exploited.

STATUS **Constitutional Monarchy**

AREA **430 sq km**

CAPITAL **Bridgetown**

POPULATION **255,000**
Density **593 people per sq km**
Life Expectancy **76 years**
Infant Mortality **10 per 1000**

LANGUAGES **English**

RELIGIONS **Protestant**

CURRENCY **Dollar**

GNP **6,370 US $ per person**

MAIN PRODUCTS **Sugar, Chemicals, Electrical components, Rum**

NATIONAL DAY **November 30**

ORGANISATIONS **UN, OAS, CARICOM, C**

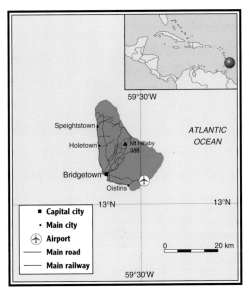

UK **COLONY** since 1652; independent 1966. Easternmost island of the Lesser Antilles chain in the Caribbean Sea. Gently rolling landscape is lush, fertile and hot, with moderate rainfall. Sugar and its by-products, molasses and rum, are important to the economy. Tourism has become a growth industry in recent years.

BELGIUM

STATUS **Constitutional Monarchy**	LANGUAGES **French, Dutch**
AREA **30,520 sq km**	RELIGIONS **Roman Catholic**
CAPITAL **Brussels**	CURRENCY **Franc**
MAIN CITIES **Antwerp, Gent, Charleroi, Liège, Bruges**	GNP **16,390 US $ per person**
POPULATION **9,845,000**	MAIN PRODUCTS **Transport equipment, Plastics, Iron and steel, Food**
Density **323 people per sq km**	NATIONAL DAY **July 21**
Life Expectancy **76 years**	ORGANISATIONS **UN, EC, OECD, NATO**
Infant Mortality **8 per 1000**	

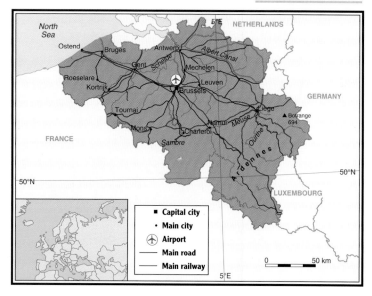

INDEPENDENT from The Netherlands 1830. Stretches from a fertile plateau in the north to the forested mountains of the Ardennes in the south. Climate is mild and temperate, the weather changeable and rainfall frequent. Cereals, root crops, vegetables and flax are the main crops. Self-sufficient in meat and dairy products. Lacks mineral resources except for coal. Metal and engineering industries are important. Brussels is the headquarters of the Commission of the EC.

BELIZE

STATUS **Constitutional Monarchy**

AREA **22,965 sq km**

CAPITAL **Belmopan**

MAIN CITIES **Belize City**

POPULATION **188,000**
Density **8 people per sq km**
Life Expectancy **67 years**
Infant Mortality **36 per 1000**

LANGUAGES **English**

RELIGIONS **Roman Catholic**

CURRENCY **Dollar**

GNP **1,600 US $ per person**

MAIN PRODUCTS **Timber, Sugar, Orange concentrate, Bananas**

NATIONAL DAY **September 21**

ORGANISATIONS **UN, CARICOM, C**

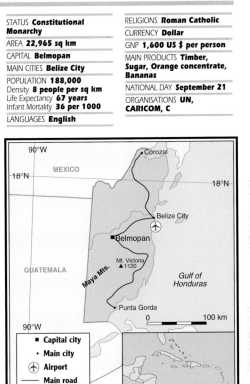

FORMER UK colony of British Honduras, renamed 1973; independent 1981. Bordering the Caribbean Sea, subtropical Belize is dominated by dense forest cover. Principal crops for export are sugar-cane, fruit, rice, maize and timber products. Since independence, has developed agriculture to reduce reliance on imported food products.

BELORUSSIA

STATUS **Republic**

AREA **208,000 sq km**

CAPITAL **Minsk**

MAIN CITIES **Gomel', Vitebsk, Mogilev, Bobruysk**

POPULATION **10,278,000**
Density **49 people per sq km**
Life Expectancy **72 years**
Infant Mortality **12 per 1000**

LANGUAGES **Belorussian**

RELIGIONS **Orthodox**

CURRENCY **Rouble**

MAIN PRODUCTS **Processed food, Chemicals, Machinery, Textiles**

ORGANISATIONS **UN**

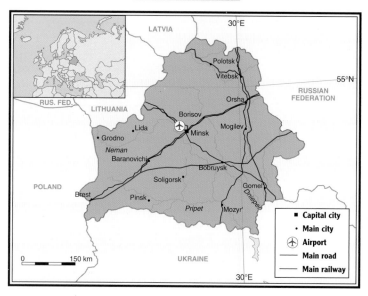

INDEPENDENT 1991. Swamps and marshlands cover large areas but, when drained, the soil is very fertile. Climate varies from maritime to continental with mild winters and high humidity. Grain, flax, potatoes and sugar beet are the main crops; livestock production very important. Most people live in central Belorussia. Comparatively poor in mineral resources. Imported raw materials and semi-manufactured goods are used in the production of trucks, agricultural machinery and other heavy engineering equipment.

STATUS **Republic**	RELIGIONS **Traditional**
AREA **112,620 sq km**	CURRENCY **CFA Franc**
CAPITAL **Porto-Novo (de facto)**	GNP **380 US $ per person**
MAIN CITIES **Cotonou (official capital)**	MAIN PRODUCTS **Cotton, Palm oil**
POPULATION **4,736,000**	NATIONAL DAY **November 30**
Density **42 people per sq km**	ORGANISATIONS **UN, OAU, ECOWAS**
Life Expectancy **48 years**	
Infant Mortality **85 per 1000**	
LANGUAGES **French**	

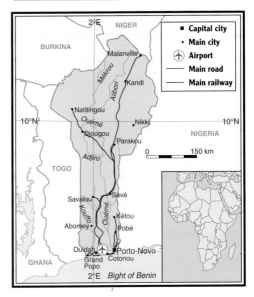

PART OF former province of French West Africa; independent 1960; name changed from Dahomey 1975. Wooded, savannah hills in the north descend to forested, cultivated lowlands fringing the Bight of Benin. Economy dominated by agriculture, with palm oil, cotton, coffee, groundnuts and copra as main exports. Developing off-shore oil industry has proven reserves of over 20 million barrels.

BERMUDA see page 218

BHUTAN

STATUS **Constitutional Monarchy**

AREA **46,620 sq km**

CAPITAL **Thimphu**

POPULATION **1,517,000**
Density **33 people per sq km**
Life Expectancy **50 years**
Infant Mortality **118 per 1000**

LANGUAGES **Dzongkha**

RELIGIONS **Buddhist**

CURRENCY **Indian Rupee, Ngultrum**

GNP **190 US $ per person**

MAIN PRODUCTS **Electricity, Timber, Minerals, Fruit and vegetables**

NATIONAL DAY **December 17**

ORGANISATIONS **UN, CP**

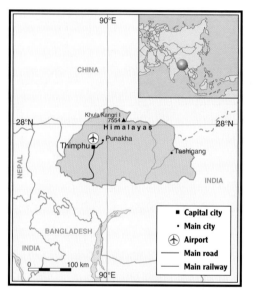

SPREADS ACROSS the Himalayan foothills between China and India east of Nepal. Rainfall is high but temperatures vary between extreme cold in the northern ranges to fairly hot in the southern forests. Economy is dominated by agriculture and small local industries. All manufactured goods are imported.

BOLIVIA

STATUS **Republic**	RELIGIONS **Roman Catholic**
AREA **1,098,575 sq km**	CURRENCY **Boliviano**
CAPITAL **La Paz**	GNP **600 US $ per person**
MAIN CITIES **Sucre (judicial capital), Santa Cruz**	MAIN PRODUCTS **Natural gas, Zinc, Tin, Silver, Soybeans**
POPULATION **7,400,000** Density **7 people per sq km** Life Expectancy **56 years** Infant Mortality **93 per 1000**	NATIONAL DAY **August 6**
	ORGANISATIONS **UN, OAS, ALADI**
LANGUAGES **Spanish**	

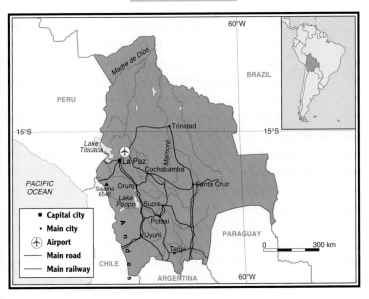

INDEPENDENT from Spain 1825. Landlocked and isolated, Bolivia stretches from the eastern Andes across high, cool plateaux before dropping to the dense forest of the Amazon basin and the grasslands of the south-east. One of the world's poorest nations. Subsistence agriculture occupies most people. Crude oil, natural gas, tin, zinc and iron ore are the main mineral deposits, the growing exploitation of which is essential for economic development.

BOSNIA-HERZEGOVINA

STATUS **Republic**

AREA **51,130 sq km**

CAPITAL **Sarajevo**

MAIN CITIES **Banja Luka, Mostar, Zenica**

POPULATION **4,200,000**
Density **82 people per sq km**
Life Expectancy **72 years**
Infant Mortality **15 per 1000**

LANGUAGES **Serbo-Croat**

RELIGIONS **Muslim, Orthodox, Roman Catholic**

CURRENCY **Dinar**

MAIN PRODUCTS **Machinery, Coal, Iron ore, Bauxite**

ORGANISATIONS **UN**

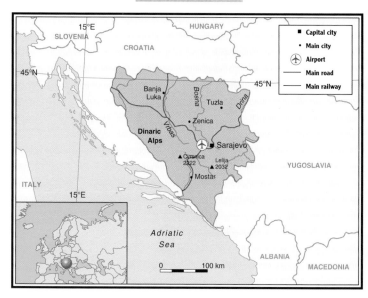

SEPARATED from Yugoslavia 1991. Virtually landlocked country with only tiny stretch of coastline. The climate near the coast is Mediterranean with hot summers, but in the mountains it is much colder. Bosnia-Herzegovina is the most ethnically mixed of the former Yugoslav republics, with the population divided into Muslims, Serbs and Croats, with no one group dominating. The struggle in Bosnia-Herzegovina has been the most bitter in the conflict following the break up of Yugoslavia.

BOTSWANA

STATUS **Republic**
AREA **600,372 sq km**
CAPITAL **Gaborone**
MAIN CITIES **Mahalapye**
POPULATION **1,291,000**
Density **2 people per sq km**
Life Expectancy **61 years**
Infant Mortality **58 per 1000**
LANGUAGES **English, Tswana**

RELIGIONS **Traditional**
CURRENCY **Pula**
GNP **940 US $ per person**
MAIN PRODUCTS **Diamonds, Copper-Nickel matte, Meat and meat products**
NATIONAL DAY **September 30**
ORGANISATIONS **UN, OAU, SADC, C**

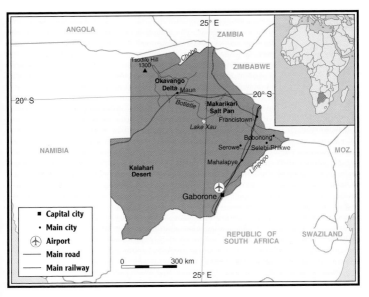

FORMER UK protectorate of Bechuanaland; independent 1966. Arid, high plateau, with poor soils and low rainfall, supports little arable agriculture, but vast numbers of cattle graze the dry grasslands. Diamonds, copper, nickel and gold are mined in the east and are the main mineral exports. Growth of light industries around the capital has stimulated trade with neighbouring countries.

BRAZIL

STATUS **Federal Republic**

AREA **8,511,965 sq km**

CAPITAL **Brasília**

MAIN CITIES **São Paulo, Rio de Janeiro, Belo Horizonte, Pôrto Alegre**

POPULATION **150,368,000**
Density **18 people per sq km**
Life Expectancy **66 years**
Infant Mortality **57 per 1000**

LANGUAGES **Portuguese**

RELIGIONS **Roman Catholic**

CURRENCY **Cruzeiro**

GNP **2,550 US $ per person**

MAIN PRODUCTS **Soya products, Boiler equipment, Iron ore and steel products**

NATIONAL DAY **September 7**

ORGANISATIONS **UN, OAS, ALADI**

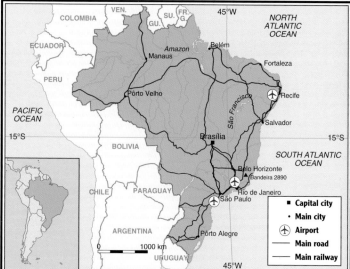

INDEPENDENT from Portugal 1822. Largest country in South America and with fastest-growing economy. Varied landscape of tropical rainforest, savanna grasslands and light forest is dominated by three river systems of the Amazon, São Francisco and Paraguay/Paraná. Major exporter of coffee, soya beans, sugar, bananas, cocoa, tobacco, rice and cattle. Mineral resources, except for iron ore, are small. Expansion of road and rail communications, light and heavy industry and hydro-electric power provides basis for industrial growth.

BRUNEI

STATUS **Monarchy**	RELIGIONS **Muslim**
AREA **5,765 sq km**	CURRENCY **Dollar**
CAPITAL **Bandar Seri Begawan**	GNP **14,120 US $ per person**
POPULATION **266,000** Density **46 people per sq km** Life Expectancy **74 years** Infant Mortality **9 per 1000**	MAIN PRODUCTS **Natural gas, Crude oil and petroleum products**
	NATIONAL DAY **February 23**
LANGUAGES **Malay**	ORGANISATIONS **UN, ASEAN, C**

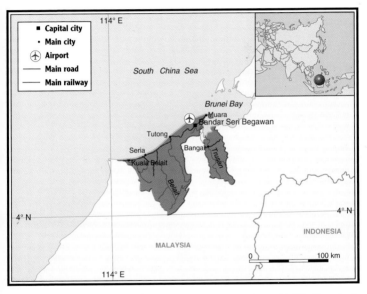

UK PROTECTORATE since 1888; independent 1984. Situated on the north-west coast of Borneo. Tropical climate is hot and humid with very high rainfall in the mountainous interior. Oil, both on-shore and off-shore, dominates the economy. Other exports include natural gas, which is transported to Japan, rubber and timber. Apart from oil, most other industries are local.

BRITISH VIRGIN ISLANDS see page 217

BULGARIA

STATUS **Republic**

AREA **110,910 sq km**

CAPITAL **Sofia**

MAIN CITIES **Plovdiv, Varna, Burgas, Ruse**

POPULATION **8,980,000**
Density **81 people per sq km**
Life Expectancy **73 years**
Infant Mortality **14 per 1000**

LANGUAGES **Bulgarian**

RELIGIONS **Orthodox**

CURRENCY **Lev**

GNP **2,320 US $ per person**

MAIN PRODUCTS **Engineering goods, Consumer goods, Food and beverages**

NATIONAL DAY **March 3**

ORGANISATIONS **UN**

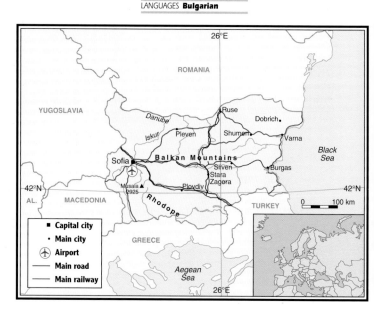

INDEPENDENT from Turkey 1908. Fertile Danube plain in the north and mountainous central and southern areas give way to the Black Sea coastal region in the east. Climate is continental with hot summers and cold winters. Black Sea resorts have considerable potential for development, as do the production and export of tobacco and wine. However, the political institutions of post-communist Bulgaria are still in a state of flux and society is undergoing a prolonged and painful crisis of transformation.

STATUS **Republic**	RELIGIONS **Traditional, Muslim**
AREA **274,200 sq km**	CURRENCY **CFA Franc**
CAPITAL **Ouagadougou**	GNP **310 US $ per person**
MAIN CITIES **Bobo-Dioulasso**	MAIN PRODUCTS **Cotton**
POPULATION **9,001,000**	NATIONAL DAY **August 4**
Density **33 people per sq km**	ORGANISATIONS **UN, OAU, ECOWAS**
Life Expectancy **49 years**	
Infant Mortality **127 per 1000**	
LANGUAGES **French**	

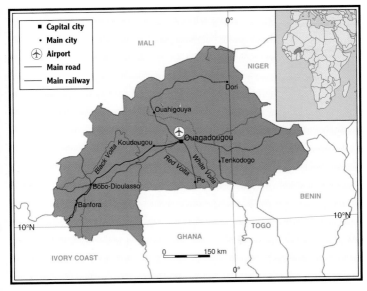

ANNEXED by France 1896; independent 1960; name changed from Upper Volta 1984. Poor, landlocked country with thin soils supporting savannah grasslands on the southern edge of the Sahara. Frequent droughts, particularly in the north, seriously affect the economy, which is mainly subsistence agriculture. Cattle and cotton are principal exports. There is virtually no industry.

BURMA (MYANMA)

STATUS **Military Regime**	LANGUAGES **Burmese**
AREA **678,030 sq km**	RELIGIONS **Buddhist**
CAPITAL **Rangoon (Yangon)**	CURRENCY **Kyat**
MAIN CITIES **Mandalay, Moulmein, Pegu, Bassein, Sittwe**	GNP **200 US $ per person**
	MAIN PRODUCTS **Agricultural products, Forest products, Minerals**
POPULATION **41,675,000** Density **61 people per sq km** Life Expectancy **63 years** Infant Mortality **59 per 1000**	NATIONAL DAY **January 4**
	ORGANISATIONS **UN, CP**

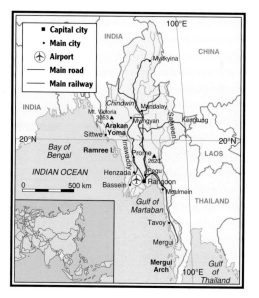

INDEPENDENT from UK 1948. Tropical rainforest predominates but is divided by the central valley of the Irrawaddy, Sittang and Salween rivers. Irrigated central basin and coastal region to the east of the Irrawaddy Delta are main rice-growing areas. Hardwoods particularly teak cover the highlands in the west. Economy is based on the export of rice and forestry products. Tin, copper, oil and natural gas deposits could be further exploited. The small amount of industry is concentrated on food processing.

BURUNDI

STATUS **Republic**

AREA **27,834 sq km**

CAPITAL **Bujumbura**

MAIN CITIES **Gitega**

POPULATION **5,458,000**
Density **196 people per sq km**
Life Expectancy **50 years**
Infant Mortality **110 per 1000**

LANGUAGES **French, Kirundi**

RELIGIONS **Roman Catholic**

CURRENCY **Franc**

GNP **220 US $ per person**

MAIN PRODUCTS **Coffee, Tea, Animal hides**

NATIONAL DAY **July 1**

ORGANISATIONS **UN, OAU, PTA**

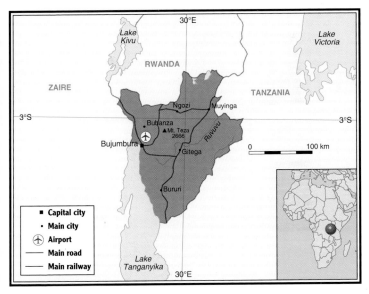

FORMER Belgian trusteeship under UN; independent 1962. One of the world's poorest nations. Poverty has two basic causes: repeated droughts and slow recovery from tribal conflicts. Situated close to the equator, yet temperatures are not high because of Burundi's altitude. Population barely produces enough food for itself. Manufacturing industry is almost non-existent.

CAMBODIA

STATUS **Republic**	RELIGIONS **Buddhist**
AREA **181,000 sq km**	CURRENCY **Riel**
CAPITAL **Phnom Penh**	GNP **130 US $ per person**
MAIN CITIES **Kompong Cham, Battambang**	MAIN PRODUCTS **Rubber, Miscellaneous manufacturing**
POPULATION **8,246,000**	NATIONAL DAY **January 7, April 17**
Density **46 people per sq km**	
Life Expectancy **51 years**	ORGANISATIONS **UN, CP**
Infant Mortality **116 per 1000**	
LANGUAGES **Khmer**	

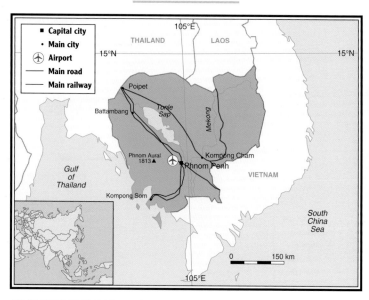

INDEPENDENT from France 1953. Potentially rich country in south-east Asia whose economy has been damaged since the 1970s by the aftermath of the Vietnam War. More than half of Cambodia is covered by monsoon rainforest. Central plain of the Mekong River covers a vast area and provides ideal conditions for rice production and harvesting of fish.

CAMEROON

STATUS **Republic**

AREA **475,500 sq km**

CAPITAL **Yaoundé**

MAIN CITIES **Douala, Garoua, Maroua**

POPULATION **11,834,000**
Density **25 people per sq km**
Life Expectancy **55 years**
Infant Mortality **86 per 1000**

LANGUAGES **French, English**

RELIGIONS **Roman Catholic**

CURRENCY **CFA Franc**

GNP **1,010 US $ per person**

MAIN PRODUCTS **Crude oil, Cocoa, Coffee, Timber, Bauxite**

NATIONAL DAY **May 20**

ORGANISATIONS **UN, OAU, UDEAC**

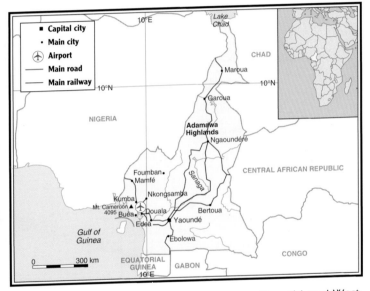

FORMED by the Union of East Cameroun (French) and West Cameroon (British) 1961. Situated just north of the equator. Coastal lowlands rise to densely forested plateaux. Rainfall varies from very high to low. Most people are farmers, and agricultural products are important export earners. Coffee and cocoa are the principal cash crops. Mineral resources are underdeveloped. One of Africa's major producers of bauxite (aluminium ore). Oil exploitation is becoming increasingly important in the economy.

STATUS	**Federal Parliamentary Monarchy**
AREA	**9,970,610 sq km**
CAPITAL	**Ottawa**
MAIN CITIES	**Toronto, Montreal, Vancouver**
POPULATION	**27,561,700**
Density	**3 people per sq km**
Life Expectancy	**77 years**
Infant Mortality	**7 per 1000**
LANGUAGES	**English, French**
RELIGIONS	**Roman Catholic, Protestant, Eastern Orthodox**
CURRENCY	**Dollar**
GNP	**19,228 US $ per person**
MAIN PRODUCTS	**Motor vehicles and parts, Grains and other foods, Primary metals, Wood pulp and paper, Oil, Gas, Timber**
NATIONAL DAY	**July 1**
ORGANISATIONS	**UN, OAS, OECD, NATO, CP, C**

Map legend:
- ■ Capital city
- • Main city
- ✈ Airport
- ─ Main road
- ─ Main railway

Bering Sea

U.S.A.

60°N

PACIFIC OCEAN

Prince Rupert

Vancouver Island

Vancou

40°N

120°

CREATED by British North American Act 1867. World's second largest country stretching from great barren islands in the Arctic north to vast grasslands in the central south, and from the Rocky Mountain chain in the west to farmlands around the Great Lakes in the east. Agriculture and mineral exploitation are very important. Oil and natural gas, iron ore, bauxite, nickel, zinc, copper, gold and silver are the major mineral exports. Despite economic success, Canada remains one of the world's most underexploited countries.

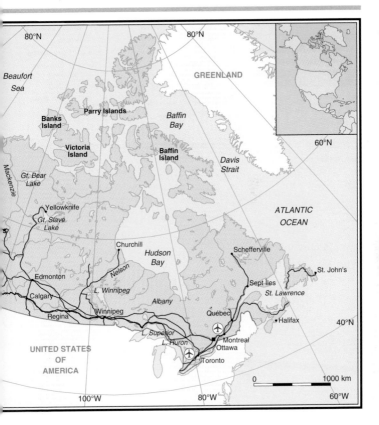

CAPE VERDE

STATUS **Republic**	RELIGIONS **Roman Catholic**
AREA **4,035 sq km**	CURRENCY **Escudo**
CAPITAL **Praia**	GNP **760 US $ per person**
MAIN CITIES **Mindelo**	MAIN PRODUCTS **Bananas, Tuna, Lobster, Sugar, Salt**
POPULATION **370,000** Density **92 people per sq km** Life Expectancy **68 years** Infant Mortality **37 per 1000**	NATIONAL DAY **July 5**
	ORGANISATIONS **UN, OAU, ECOWAS**
LANGUAGES **Portuguese**	

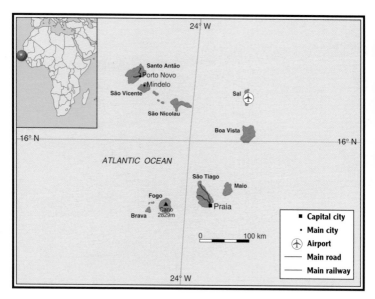

F ORMER Portuguese overseas province; independent 1975. Comprises ten inhabited volcanic islands situated in the Atlantic Ocean off Senegal. Rainfall is low but irrigation encourages growth of sugar-cane, coconuts, fruit and maize. Fishing makes major contribution to export revenue. All consumer goods are imported and trading links continue to be maintained with Portugal.

CAYMAN ISLANDS see page 217

CENTRAL AFRICAN REPUBLIC

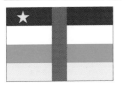

STATUS **Republic**

AREA **624,975 sq km**

CAPITAL **Bangui**

POPULATION **3,039,000**
Density **5 people per sq km**
Life Expectancy **51 years**
Infant Mortality **95 per 1000**

LANGUAGES **French, Sango**

RELIGIONS **Traditional**

CURRENCY **CFA Franc**

GNP **390 US $ per person**

MAIN PRODUCTS **Coffee, Diamonds, Timber, Tobacco, Cotton**

NATIONAL DAY **December 1**

ORGANISATIONS **UN, OAU, UDEAC**

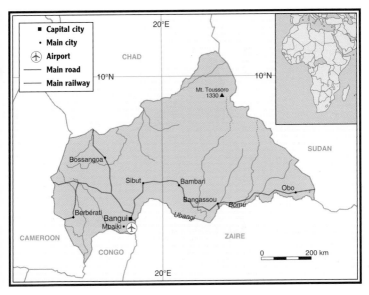

PART OF former French Equatorial Africa; independent 1960. Landlocked and remote. Tropical climate with little variation in temperature. Savannah covers the rolling plateaux with rainforest in the south-east. To the north lies the Sahara Desert. Most farming is at subsistence level with a small amount of crops grown for export - cotton, coffee and tobacco. Diamonds and uranium ore are the major mineral exports along with some gold. Hardwood forests in the south-west provide timber for export.

CHAD

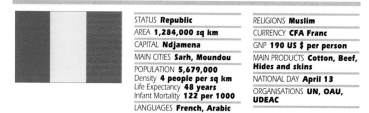

STATUS **Republic**	RELIGIONS **Muslim**
AREA **1,284,000 sq km**	CURRENCY **CFA Franc**
CAPITAL **Ndjamena**	GNP **190 US $ per person**
MAIN CITIES **Sarh, Moundou**	MAIN PRODUCTS **Cotton, Beef, Hides and skins**
POPULATION **5,679,000**	
Density **4 people per sq km**	NATIONAL DAY **April 13**
Life Expectancy **48 years**	ORGANISATIONS **UN, OAU, UDEAC**
Infant Mortality **122 per 1000**	
LANGUAGES **French, Arabic**	

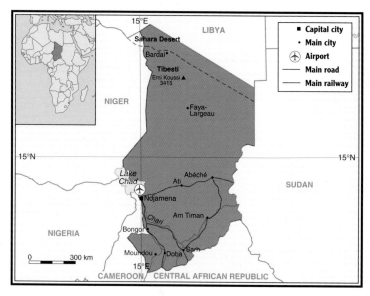

PART OF former French Equatorial Africa; independent 1960. One of the world's poorest countries. Stretches deep into the Sahara Desert. Economy based on agriculture, but only the south can support crops and animals. Severe droughts, increasing desertification and border disputes severely restrict development. Salt is mined around Lake Chad, where most people live.

STATUS **Republic**

AREA **751,625 sq km**

CAPITAL **Santiago**

MAIN CITIES **Valapraiso, Concepción, Antofagasta**

POPULATION **13,173,000**
Density **18 people per sq km**
Life Expectancy **72 years**
Infant Mortality **19 per 1000**

LANGUAGES **Spanish**

RELIGIONS **Roman Catholic**

CURRENCY **Peso**

GNP **1,770 US $ per person**

MAIN PRODUCTS **Copper and other minerals, Paper, Chemicals, Fruit and vegetables**

NATIONAL DAY **September 18**

ORGANISATIONS **UN, OAS, ALADI**

INDEPENDENT from Spain 1818. Long, thin country on west coast of South America. Subtropical deserts in the north; ice deserts in the south. Apart from a narrow coastal strip of lowland, Chile is dominated by the Andean mountains. Economy based on abundant mineral resources, with copper (the world's largest reserve), iron ore, nitrates, coal, all major exports. Most energy is provided by hydro-electric power. Light and heavy industries are based around Concepción and Santiago.

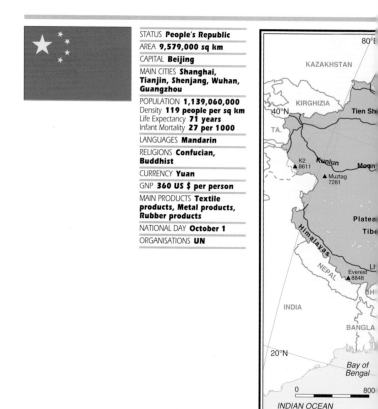

STATUS	**People's Republic**
AREA	**9,579,000 sq km**
CAPITAL	**Beijing**
MAIN CITIES	**Shanghai, Tianjin, Shenjang, Wuhan, Guangzhou**
POPULATION	**1,139,060,000**
	Density **119 people per sq km**
	Life Expectancy **71 years**
	Infant Mortality **27 per 1000**
LANGUAGES	**Mandarin**
RELIGIONS	**Confucian, Buddhist**
CURRENCY	**Yuan**
GNP	**360 US $ per person**
MAIN PRODUCTS	**Textile products, Metal products, Rubber products**
NATIONAL DAY	**October 1**
ORGANISATIONS	**UN**

COUNTRY with the largest population, one in five of the world's population live in China. Most people live in the east, where intensive irrigated agriculture produces rice, wheat, maize, sugar, soya beans and oil seeds. Self-sufficient in cereals, fish and livestock. Most of the north is desert. Vast underexploited reserves of tin, copper, zinc, coal, iron ore and oil form basis for industrial growth. Trade with USA, western Europe and Japan is increasing.

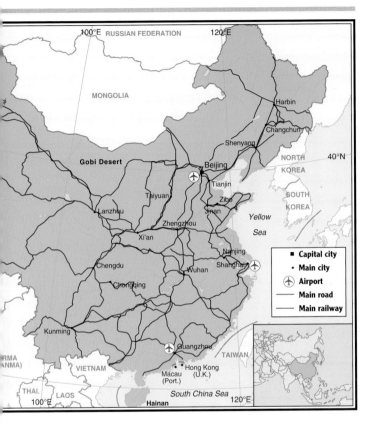

Legend:
- ■ Capital city
- • Main city
- ✈ Airport
- — Main road
- — Main railway

Map labels:
- RUSSIAN FEDERATION
- MONGOLIA
- Gobi Desert
- Harbin
- Changchun
- Shenyang
- NORTH KOREA
- SOUTH KOREA
- Beijing
- Tianjin
- Taiyuan
- Zibo
- Jinan
- Lanzhou
- Zhengzhou
- Xi'an
- Yellow Sea
- Chengdu
- Wuhan
- Nanjing
- Shanghai
- Chongqing
- Kunming
- Guangzhou
- TAIWAN
- Macau (Port.)
- Hong Kong (U.K.)
- VIETNAM
- BURMA (MYANMA)
- THAI.
- LAOS
- South China Sea
- Hainan
- 100°E
- 120°E
- 40°N

CHRISTMAS ISLAND see page 36-37
COCOS ISLANDS see page 36-37

COLOMBIA

STATUS **Republic**

AREA **1,138,915 sq km**

CAPITAL **Bogotá**

MAIN CITIES **Medellin, Cali, Barranquila, Cartagena, Cúcuta**

POPULATION **32,987,000**
Density **29 people per sq km**
Life Expectancy **69 years**
Infant Mortality **37 per 1000**

LANGUAGES **Spanish**

RELIGIONS **Roman Catholic**

CURRENCY **Peso**

GNP **1,190 US $ per person**

MAIN PRODUCTS **Minerals, Crude oil, Coffee, Clothing, Fruit**

NATIONAL DAY **July 20**

ORGANISATIONS **UN, OAS, ALADI**

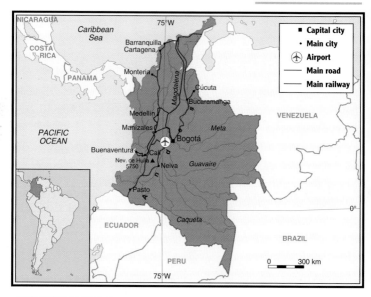

INDEPENDENT from Spain 1819. Bounded by sea in the north and west, the Andean mountains run from north to south. Sources of Amazon and Orinoco rivers in the east. Most of country covered by tropical rainforest. Fertile river valleys in the uplands produce coffee - a major export. Bananas, tobacco, cotton, sugar and rice grow at lower altitudes. Manufacturing as well as mining of coal, iron ore, copper and precious stones becoming more dominant in the economy. Immense illegal quantities of cocaine are exported to USA and elsewhere.

STATUS **Federal Islamic Republic**	RELIGIONS **Muslim**
AREA **1,860 sq km**	CURRENCY **CFA Franc**
CAPITAL **Moroni**	GNP **460 US $ per person**
POPULATION **551,000**	MAIN PRODUCTS **Vanilla, Ylang-Ylang, Cloves**
Density **296 people per sq km**	
Life Expectancy **56 years**	NATIONAL DAY **July 6**
Infant Mortality **89 per 1000**	ORGANISATIONS **UN, OAU, PTA**
LANGUAGES **Arabic, French**	

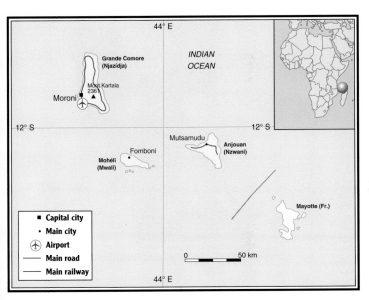

FORMER French overseas territory; independent 1975. Comprises three islands situated between Madagascar and east African coast. Cool, dry season alternates with hot, humid monsoon one; rainfall is moderate. Mangoes, coconuts and bananas grow around the coastal lowlands. Economy based on export of vanilla, ylang-ylang and cloves. Timber and timber products are important to local development. There is no manufacturing.

CONGO

STATUS **Republic**	RELIGIONS **Traditional**
AREA **342,000 sq km**	CURRENCY **CFA Franc**
CAPITAL **Brazzaville**	GNP **930 US $ per person**
MAIN CITIES **Pointe Noire**	MAIN PRODUCTS **Crude oil and petroleum products, Timber, Diamonds**
POPULATION **2,271,000**	
Density **7 people per sq km**	
Life Expectancy **55 years**	NATIONAL DAY **August 15**
Infant Mortality **65 per 1000**	ORGANISATIONS **UN, OAU, UDEAC**
LANGUAGES **French**	

PART OF former French Equatorial Africa; independent 1960. Africa's first communist state. Retains strong economic ties with the West, especially France. Mostly swamp and forest; wooded savannah on the upland Batéké plateau in the west. Many people subsistence farm, growing plantains, maize and cassava. Coffee, groundnuts, cocoa, timber and timber products are exported. Considerable reserves of industrial diamonds, gold, lead, zinc and oil.

COOK ISLANDS see page 157

COSTA RICA

STATUS **Republic**	RELIGIONS **Roman Catholic**
AREA **50,900 sq km**	CURRENCY **Colón**
CAPITAL **San José**	GNP **1,790 US $ per person**
MAIN CITIES **Limón**	MAIN PRODUCTS **Coffee,**
POPULATION **2,994,000**	**Bananas, Textiles, Beef, Fish**
Density **59 people per sq km**	NATIONAL DAY **September 15**
Life Expectancy **75 years**	ORGANISATIONS **UN, OAS,**
Infant Mortality **17 per 1000**	**CACM**
LANGUAGES **Spanish**	

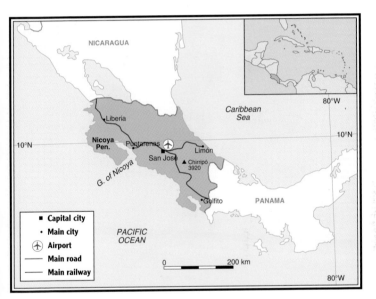

SPANISH colony since 1530; independent 1821. Has coastlines with both the Pacific Ocean and Caribbean Sea. Mountain chains running the length of this narrow country form fertile uplands where coffee, one of the main crops and exports, and cattle flourish. Bananas grow on the coast of the Pacific Ocean. Gold, silver, iron ore and bauxite are mined. Principal industries are food processing and manufacture of textiles and chemicals, fertilizers and furniture.

CROATIA

STATUS **Republic**

AREA **56,540 sq km**

CAPITAL **Zagreb**

MAIN CITIES **Osijek, Rijeka, Split**

POPULATION **4,600,000**
Density **81 people per sq km**
Life Expectancy **72 years**
Infant Mortality **10 per 1000**

LANGUAGES **Serbo-Croat**

RELIGIONS **Roman Catholic**

CURRENCY **Dinar**

MAIN PRODUCTS **Crude oil, Machinery, Cement**

ORGANISATIONS **UN**

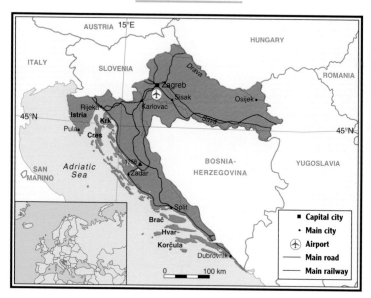

SEPARATED from Yugoslavia in 1991. Runs in a narrow strip along the Adriatic Sea, then extends inland in a broad curve. Intensively farmed fertile plains in centre and east provide surplus crops, meat and dairy products. Mountainous, barren coastal regions have been developed for tourism. Economy severely harmed by military conflict in other parts of former Yugoslavia. Prosperity from electrical engineering, metalworking and machine-building, chemicals and rubber industries could be jeopardised. Tourism has all but collapsed.

STATUS **Socialist People's Republic**	LANGUAGES **Spanish**
AREA **114,525 sq km**	RELIGIONS **Roman Catholic**
CAPITAL **Havana**	CURRENCY **Peso**
MAIN CITIES **Santiago de Cuba, Camagüey, Guantánamo, Holguín**	GNP **2,000 US $ per person**
	MAIN PRODUCTS **Sugar, Minerals, Citrus fruit, Fish, Tobacco**
POPULATION **10,609,000** Density **93 people per sq km** Life Expectancy **76 years** Infant Mortality **13 per 1000**	NATIONAL DAY **January 1**
	ORGANISATIONS **UN**

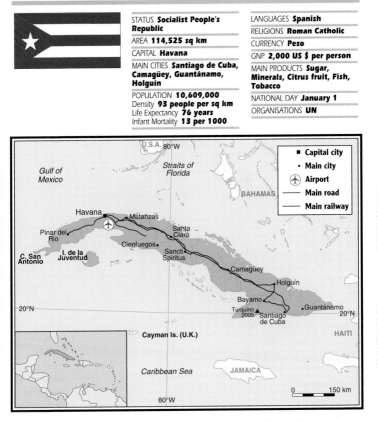

INDEPENDENT from Spain 1898. Consists of one large island and over 1,500 small ones. Varied landscape of fertile plains, mountain ranges and gentle countryside, with moderate temperatures and rainfall. Being the only communist state in the Americas, most trading is done with former USSR and former Comecon countries. Sugar, tobacco and nickel are main exports. Mining of manganese, chrome, copper and oil is expanding. Cuba has enough cattle and coffee for domestic use but many other food products are imported.

CYPRUS

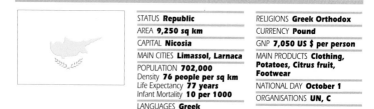

STATUS **Republic**	RELIGIONS **Greek Orthodox**
AREA **9,250 sq km**	CURRENCY **Pound**
CAPITAL **Nicosia**	GNP **7,050 US $ per person**
MAIN CITIES **Limassol, Larnaca**	MAIN PRODUCTS **Clothing, Potatoes, Citrus fruit, Footwear**
POPULATION **702,000** Density **76 people per sq km** Life Expectancy **77 years** Infant Mortality **10 per 1000**	NATIONAL DAY **October 1**
	ORGANISATIONS **UN, C**
LANGUAGES **Greek**	

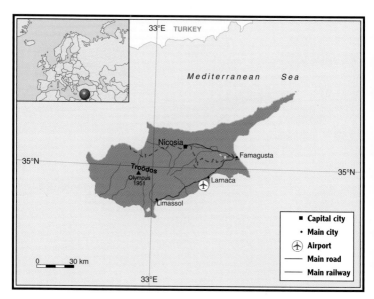

UK COLONY since 1925; independent 1960. Prosperous island in the Mediterranean Sea. Summers are very hot and dry; winters warm and wet. Most of island is under cultivation, producing citrus fruit, potatoes, barley, wheat and olives. Sheep, goats and pigs are the principal livestock. Tourism is an important source of foreign exchange, despite Turkish occupation of the north. Most industry consists of local manufacturing.

STATUS **Republic**	RELIGIONS **Roman Catholic**
AREA **78,863 sq km**	CURRENCY **Koruna**
CAPITAL **Prague**	GNP **3,140 US $ per person**
MAIN CITIES **Brno, Ostrava, Plzeň, Olomouc**	MAIN PRODUCTS **Machinery, Transport equipment, Consumer goods**
POPULATION **10,356,000** Density **131 people per sq km** Life Expectancy **72 years** Infant Mortality **11 per 1000**	ORGANISATIONS **UN**
LANGUAGES **Czech**	

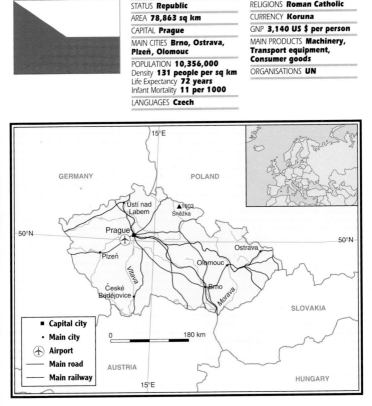

SEPARATED from Slovakia 1993. Landlocked country in the centre of Europe, consisting of rolling hills and plateaux in the west, with the land rising in the east. The climate is continental with harsh winters and warm summers; the rainfall is low. Though deficient in energy resources and raw materials, the Czech Republic is one of the wealthiest of the old eastern bloc states, but the communist legacy has hindered growth. The Czech Republic is looking to the west, especially Austria and Germany, for economic growth.

DENMARK

STATUS **Constitutional Monarchy**	RELIGIONS **Protestant**
AREA **43,075 sq km**	CURRENCY **Krone**
CAPITAL **Copenhagen**	GNP **20,510 US $ per person**
MAIN CITIES **Århus, Odense, Ålborg, Esbjerg**	MAIN PRODUCTS **Machinery and instruments, Agricultural products, Chemical products**
POPULATION **5,140,000** Density **119 people per sq km** Life Expectancy **76 years** Infant Mortality **6 per 1000**	NATIONAL DAY **April 16**
	ORGANISATIONS **UN, EC, OECD, NATO**
LANGUAGES **Danish**	

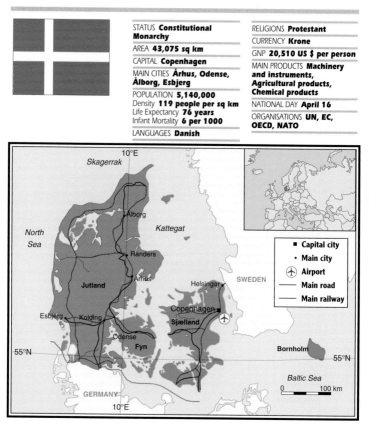

THIS 'bridge' between Germany and Scandinavia consists of the Jutland Peninsula and over 400 islands, of which only one quarter are inhabited. Climate is mild, with most rain falling in summer and autumn. Meat and dairy products - beef, butter, cheese, eggs, bacon and pork - are exported, as are manufactured goods. Extensive fishing industry is centred on shallow lagoons that have formed along indented western coastline. Recently fishing industry has had problems with over-fishing and quotas. Mineral resources are scarce.

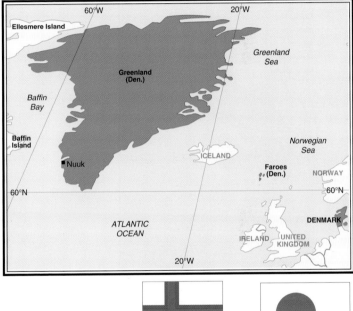

FAROES

STATUS **Self Governing part of the Danish Realm**

AREA **1,399 sq km**

CAPITAL **Tórshavn**

POPULATION **48,000**
Density **34 people per sq km**

LANGUAGES **Danish, Faroese**

GREENLAND

STATUS **Self Governing part of the Danish Realm**

AREA **2,175,600 sq km**

CAPITAL **Nuuk**

POPULATION **57,000**
Density **1 person per sq km**

LANGUAGES **Danish, Greenlandic**

DJIBOUTI

STATUS **Republic**

AREA **23,000 sq km**

CAPITAL **Djibouti**

POPULATION **409,000**
Density **18 people per sq km**
Life Expectancy **49 years**
Infant Mortality **112 per 1000**

LANGUAGES **French, Somali**

RELIGIONS **Muslim**

CURRENCY **Franc**

GNP **1,070 US $ per person**

MAIN PRODUCTS **Live animals, Food products**

NATIONAL DAY **June 27**

ORGANISATIONS **UN, OAU, PTA, AL**

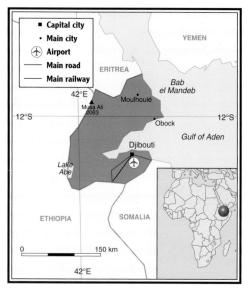

FORMER French Territory of the Afars and Issas; independent 1977. Situated at the mouth of the Red Sea. Port of Djibouti is important transit point for Red Sea trade, especially for Ethiopia. Climate is extremely hot and arid; rainfall is low. There is consequently very little cultivation.

STATUS **Republic**
AREA **751 sq km**
CAPITAL **Roseau**

POPULATION **83,000**
Density **111 people per sq km**
Life Expectancy **74 years**
Infant Mortality **14 per 1000**

LANGUAGES **English, French**

RELIGIONS **Roman Catholic**
CURRENCY **EC Dollar**
GNP **1,650 US $ per person**
MAIN PRODUCTS **Bananas, Soap, Fruit juices**
NATIONAL DAY **November 3**
ORGANISATIONS **UN, OAS, CARICOM, C**

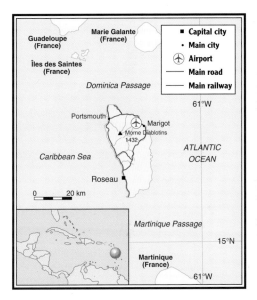

INDEPENDENT from UK 1978. Located in the Windward Islands in the east Caribbean Sea and covered by tropical rainforest. Obtains foreign revenue from sugar-cane, bananas, coconuts, soap, vegetables and citrus fruits. Tourism is the most rapidly expanding industry.

DOMINICAN REPUBLIC

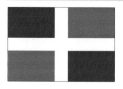

STATUS **Republic**	RELIGIONS **Roman Catholic**
AREA **48,440 sq km**	CURRENCY **Peso**
CAPITAL **Santo Domingo**	GNP **790 US $ per person**
MAIN CITIES **Santiago, La Romana**	MAIN PRODUCTS **Nickel, Sugar, Gold, Coffee**
POPULATION **7,170,000** Density **148 people per sq km** Life Expectancy **68 years** Infant Mortality **57 per 1000**	NATIONAL DAY **February 27**
	ORGANISATIONS **UN, OAS**
LANGUAGES **Spanish**	

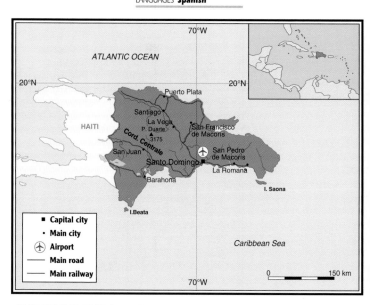

INDEPENDENT from Spain 1821. Occupies eastern part of the Caribbean island of Hispaniola. Landscape is dominated by a series of mountain ranges, thickly covered with rainforest. In the south is a coastal plain where the capital, Santo Domingo, lies. Rainfall is moderate. Economy based principally on agriculture; sugar, coffee, cocoa and tobacco are staple crops. Minerals include bauxite, nickel, gold and silver.

STATUS **Republic**	RELIGIONS **Roman Catholic**
AREA **461,475 sq km**	CURRENCY **Sucre**
CAPITAL **Quito**	GNP **1,040 US $ per person**
MAIN CITIES **Guayaquil, Cuenca, Machala, Riobamba**	MAIN PRODUCTS **Crude oil, Bananas, Shrimp, Coffee, Cocoa**
POPULATION **10,782,000** Density **23 people per sq km** Life Expectancy **67 years** Infant Mortality **57 per 1000**	NATIONAL DAY **August 10**
	ORGANISATIONS **UN, OAS, ALADI**
LANGUAGES **Spanish**	

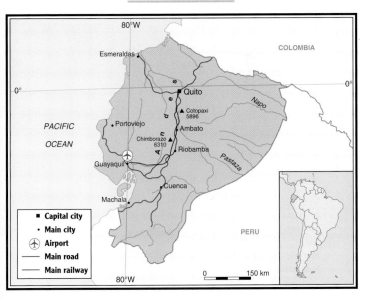

PART OF Spanish Vice-royalty of Peru from 16th century; independent 1822. Main agricultural exports - bananas, coffee and cocoa - grow in the fertile coastal lowlands. Inland, stretching north to south, are the Andean highlands; here maize, wheat and barley are cultivated. Large resources of crude oil are being exploited in the thickly forested lowlands in the east. Fishing industry, especially shrimps, is growing rapidly. Mineral reserves include silver, gold, copper and zinc.

EGYPT

STATUS **Republic**	RELIGIONS **Muslim**
AREA **1,000,250 sq km**	CURRENCY **Pound**
CAPITAL **Cairo**	GNP **630 US $ per person**
MAIN CITIES **Alexandria, El Giza, Port Said, Suez**	MAIN PRODUCTS **Cotton, Textiles, Fabrics, Crude oil and petroleum products**
POPULATION **53,153,000** Density **53 people per sq km** Life Expectancy **62 years** Infant Mortality **57 per 1000**	NATIONAL DAY **July 23**
	ORGANISATIONS **UN, OAU, AL**
LANGUAGES **Arabic**	

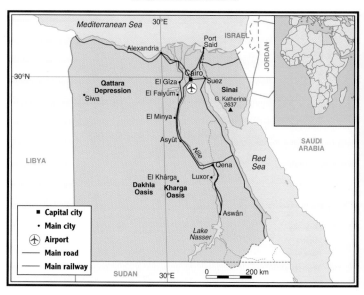

UK **PROTECTORATE** 1914-1922; republic 1953. Fertile, irrigated Nile Valley, surrounded by two deserts. Dependent on water from the Nile as rainfall varies from low in the north to non-existent in the deserts. Cotton is an important crop; also cereals, fruits, rice, sugar-cane and vegetables. Buffalo, cattle, sheep, goats and camels are principal livestock. Tourism and tolls from Suez Canal earn important revenue. Main mineral deposits are phosphates, iron ore, salt, manganese and chromium.

EL SALVADOR

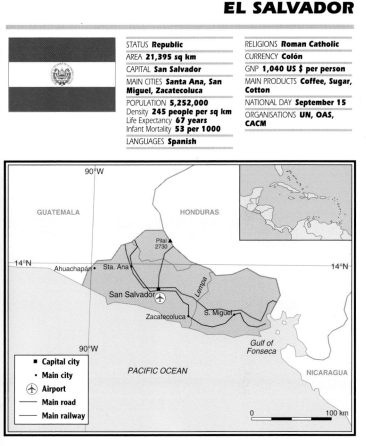

STATUS **Republic**	RELIGIONS **Roman Catholic**
AREA **21,395 sq km**	CURRENCY **Colón**
CAPITAL **San Salvador**	GNP **1,040 US $ per person**
MAIN CITIES **Santa Ana, San Miguel, Zacatecoluca**	MAIN PRODUCTS **Coffee, Sugar, Cotton**
POPULATION **5,252,000** Density **245 people per sq km** Life Expectancy **67 years** Infant Mortality **53 per 1000**	NATIONAL DAY **September 15** ORGANISATIONS **UN, OAS, CACM**
LANGUAGES **Spanish**	

Map labels: 90°W · GUATEMALA · HONDURAS · Pital ▲ 2730 · 14°N · Ahuachapán · Sta. Ana · Lempa · San Salvador · Zacatecoluca · S. Miguel · Gulf of Fonseca · 90°W · PACIFIC OCEAN · NICARAGUA

- ■ **Capital city**
- • **Main city**
- ✈ **Airport**
- — **Main road**
- — **Main railway**

0   100 km

INDEPENDENT from Spain 1821. Small, densely populated Central American country on the Pacific Ocean. Most people live around the central plain. Temperatures are tropical but not too hot due to El Salvador's altitude; rainfall is moderate. Coffee, cotton and sugar are important exports. Industry has expanded considerably with the production of textiles, shoes, cosmetics, cement, processed foods, chemicals and furniture. Mineral resources are negligible.

EQUATORIAL GUINEA

STATUS **Republic**	RELIGIONS **Roman Catholic**
AREA **28,050 sq km**	CURRENCY **CFA Franc**
CAPITAL **Malabo**	GNP **430 US $ per person**
MAIN CITIES **Bata**	MAIN PRODUCTS **Cocoa, Fuels and lubricants, Timber**
POPULATION **348,000** Density **12 people per sq km** Life Expectancy **48 years** Infant Mortality **117 per 1000**	NATIONAL DAY **October 12** ORGANISATIONS **UN, OAU, UDEAC**
LANGUAGES **Spanish**	

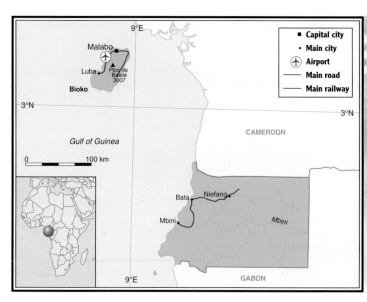

I NDEPENDENT from Spain 1968. Comprises mainland Mbini - with hot, wet climate and dense rainforest but little economic development - and the volcanic island of Bioko. Agriculture is important as are forest products, fish and processed foods manufactured near the coast in Mbini.

ERITREA

STATUS **Republic**
AREA **93,679 sq km**
CAPITAL **Asmara**
MAIN CITIES **Massawa**
POPULATION **3,500,000**
Density **37 people per sq km**
LANGUAGES **Arabic, English**

RELIGIONS **Christian, Muslim**
CURRENCY **Ethiopian Birr**
NATIONAL DAY **May 24**
ORGANISATIONS **UN**

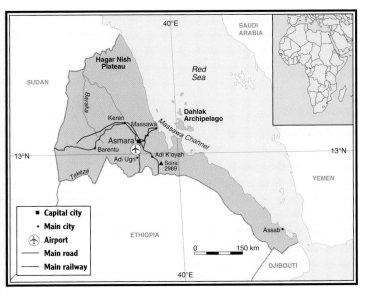

Former Italian colony handed over to Ethiopia by United Nations in 1952. Fully integrated into Ethiopia 1962. Independent in 1993 after 30 year struggle. Narrow coastal strip rises to high temperate plateaux. Most people are pastoral farmers. Red Sea ports should facilitate economic development. Secession leaves Ethiopia landlocked.

ESTONIA

STATUS **Republic**

AREA **45,100 sq km**

CAPITAL **Tallinn**

MAIN CITIES **Tartu, Narva, Kohtla-Järve, Pärnu**

POPULATION **1,583,000**
Density **35 people per sq km**
Life Expectancy **70 years**
Infant Mortality **15 per 1000**

LANGUAGES **Estonian**

RELIGIONS **Protestant**

CURRENCY **Kroon**

MAIN PRODUCTS **Textiles, Cement, Processed foods, Paper**

ORGANISATIONS **UN**

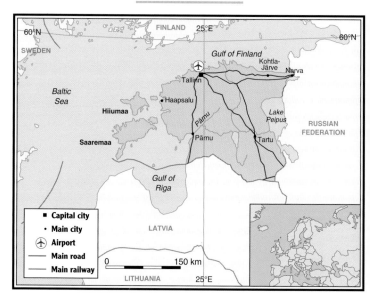

INCORPORATED into USSR 1940; independent 1991. Smallest and most northerly of the Baltic States, with many islands. Generally flat or undulating mainland, with extensive forests and many lakes, which is difficult to farm. Climate is temperate. Oil shale and phosphorite are only important natural resources. Engineering, metalworking, timber production, woodworking and textiles are principal industries. Economy is currently undergoing a profound transformation to a free market system based on private enterprise.

ETHIOPIA

STATUS **Republic**
AREA **1,128,221 sq km**
CAPITAL **Addis Ababa**
MAIN CITIES **Diredawa**
POPULATION **47,474,000**
Density **42 people per sq km**
Life Expectancy **47 years**
Infant Mortality **122 per 1000**
LANGUAGES **Amharic**

RELIGIONS **Orthodox**
CURRENCY **Birr**
GNP **120 US $ per person**
MAIN PRODUCTS **Coffee, Hides, Live animals, Pulses**
NATIONAL DAY **September 12**
ORGANISATIONS **UN, OAU, PTA**

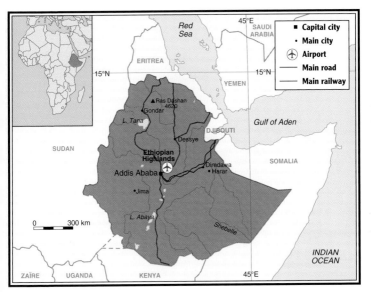

CONSISTS OF high plateaux and plains of arid desert. Farming in the high rural areas generates most export revenue. Main crops are coffee, fruit, vegetables and oil seeds. Gold and salt are mined on a small scale. The most important industries are cotton textiles, cement, canned foods, construction materials and leather goods. Difficulty in communications has hindered development as have recent droughts and civil wars. Secession of Eritrea has left Ethiopia landlocked.

FALKLAND ISLANDS see page 218
FAROES see page 77

FIJI

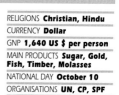

STATUS **Republic**	RELIGIONS **Christian, Hindu**
AREA **18,330 sq km**	CURRENCY **Dollar**
CAPITAL **Suva**	GNP **1,640 US $ per person**
MAIN CITIES **Lautoka**	MAIN PRODUCTS **Sugar, Gold, Fish, Timber, Molasses**
POPULATION **765,000** Density **42 people per sq km** Life Expectancy **66 years** Infant Mortality **24 per 1000**	NATIONAL DAY **October 10** ORGANISATIONS **UN, CP, SPF**
LANGUAGES **English, Fiji, Hindi**	

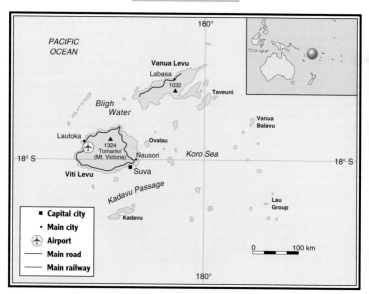

INDEPENDENT from UK 1970; republic 1987. Comprises some 320 tropical islands in the south central Pacific Ocean, of which more than 100 are inhabited. Economy is geared to production of sugar-cane, coconut oil, bananas and rice. Main industries are sugar processing, gold mining, copra processing and fish canning. Important livestock are cattle, goats, pigs and poultry.

FINLAND

STATUS **Republic**

AREA **337,030 sq km**

CAPITAL **Helsinki**

MAIN CITIES **Tampere, Turku, Vantaa, Espoo, Oulu**

POPULATION **4,986,000**
Density **15 people per sq km**
Life Expectancy **76 years**
Infant Mortality **5 per 1000**

LANGUAGES **Finnish, Swedish**

RELIGIONS **Protestant**

CURRENCY **Markka**

GNP **22,060 US $ per person**

MAIN PRODUCTS **Paper and paper products, Metal products, Timber**

NATIONAL DAY **December 6**

ORGANISATIONS **UN, EFTA, OECD**

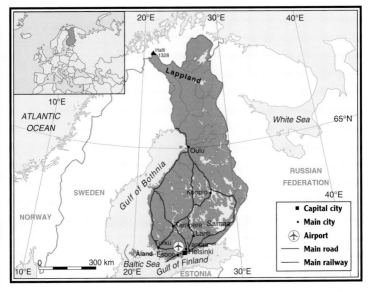

PART OF Grand Duchy of Russia since 1809; independent 1917. Flat land of lakes and forests. Soils are thin and poor on the ice-scarred, granite plateau, but most of Finland is forested with conifers. Timber and timber products make up most of Finnish exports. Because of the harsh northern climate, most people live in towns in the far south. Manufacturing industry has been developing rapidly in recent years.

FRANCE

STATUS **Republic**

AREA **543,965 sq km**

CAPITAL **Paris**

MAIN CITIES **Lyon, Marseille, Lille, Bordeaux, Toulouse, Nantes**

POPULATION **56,440,000**
Density **104 people per sq km**
Life Expectancy **77 years**
Infant Mortality **7 per 1000**

LANGUAGES **French**

RELIGIONS **Roman Catholic**

CURRENCY **Franc**

GNP **17,830 US $ per person**

MAIN PRODUCTS **Machinery, Agricultural products, Transport equipment**

NATIONAL DAY **July 14**

ORGANISATIONS **UN, EC, OECD, NATO**

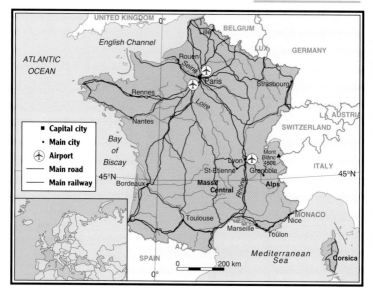

ENCOMPASSES a series of high plateaux, mountain ranges and lowland basins. Climate is generally mild: south has a Mediterranean climate with hot, dry summers; the rest of France has rain all year round. Much land is agricultural. Self-sufficient in cereals, meat, dairy products, fruit and vegetables. Leading exporter of wheat, barley, sugar beet and wine. Reserves of coal, oil and natural gas. A leading world producer of vehicles, aeronautical and refined-chemical products, armaments, fashion perfumes and luxury goods.

FRANCE Overseas Departments

GUADELOUPE

STATUS **Overseas Department**

AREA **1,705 sq km**

CAPITAL **Basse-Terre**

POPULATION **344,000**
Density **202 people per sq km**

LANGUAGES **French**

FRENCH GUIANA

STATUS **Overseas Department**

AREA **91,000 sq km**

CAPITAL **Cayenne**

POPULATION **99,000**
Density **1 person per sq km**

LANGUAGES **French**

MARTINIQUE

STATUS **Overseas Department**

AREA **1,079 sq km**

CAPITAL **Fort-de-France**

POPULATION **361,000**
Density **335 people per sq km**

LANGUAGES **French**

FRANCE Overseas Departments

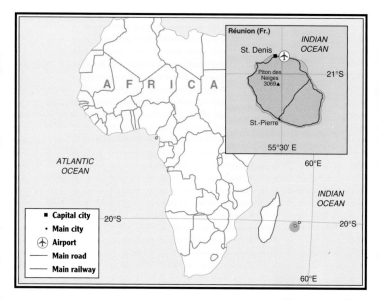

Réunion (Fr.)

INDIAN OCEAN

St. Denis

Piton des Neiges 3069▲

21°S

St.-Pierre

55°30' E

AFRICA

ATLANTIC OCEAN

60°E

INDIAN OCEAN

- ■ Capital city
- • Main city
- ⊕ Airport
- — Main road
- — Main railway

20°S   20°S

60°E

RÉUNION
STATUS **Overseas Department**
AREA **2,512 sq km**
CAPITAL **St.-Denis**
POPULATION **592,000**
Density **236 people per sq km**
LANGUAGES **French**

FRANCE Overseas Territories

FRENCH POLYNESIA
STATUS **Overseas Territory**
AREA **3,265 sq km**
CAPITAL **Papeete**
POPULATION **206,000**
Density **63 people per sq km**
LANGUAGES **French, Tahitian**

MAYOTTE
STATUS **Territorial Collectivity**
AREA **362 sq km**
CAPITAL **Mamoudzou**
POPULATION **57,000**
Density **157 people per sq km**
LANGUAGES **French**

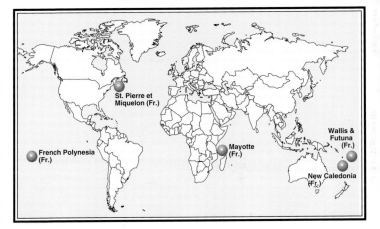

NEW CALEDONIA
STATUS **Overseas Territory**
AREA **19,058 sq km**
CAPITAL **Nouméa**
POPULATION **168,000**
Density **9 people per sq km**
LANGUAGES **French**

ST. PIERRE ET MIQUELON
STATUS **Territorial Collectivity**
AREA **242 sq km**
CAPITAL **St. Pierre**
POPULATION **7,000**
Density **29 people per sq km**
LANGUAGES **French**

WALLIS AND FUTUNA
STATUS **Overseas Territory**
AREA **274 sq km**
CAPITAL **Mata-Utu**
POPULATION **18,000**
Density **66 people per sq km**
LANGUAGES **French, Wallisian**

GABON

STATUS **Republic**	RELIGIONS **Roman Catholic**
AREA **267,665 sq km**	CURRENCY **CFA Franc**
CAPITAL **Libreville**	GNP **2,770 US $ per person**
MAIN CITIES **Port Gentil**	MAIN PRODUCTS **Crude oil and petroleum products, Manganese, Timber**
POPULATION **1,172,000** Density **4 people per sq km** Life Expectancy **54 years** Infant Mortality **94 per 1000**	NATIONAL DAY **August 17**
LANGUAGES **French**	ORGANISATIONS **UN, OAU, UDEAC, OPEC**

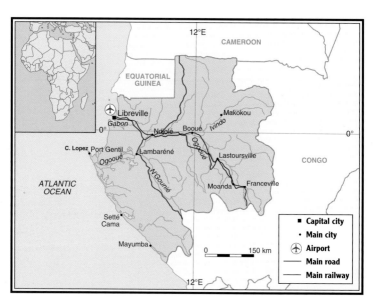

PART OF former French Equatorial Africa; independent 1960. Lies on the equator in western central Africa, mostly in basin of River Ogooué. Covered in tropical rainforest. Is hot and wet all year round. One of the most prosperous states in Africa with valuable timber and mineral resources. Main exports are oil and oil products, timber, manganese and uranium. Palm oil, rubber, coffee and cocoa are grown for export; cassava, plantain and maize for subsistence. Major reserves of iron ore are underexploited.

GAMBIA

STATUS **Republic**	RELIGIONS **Muslim**
AREA **10,690 sq km**	CURRENCY **Dalasi**
CAPITAL **Banjul**	GNP **230 US $ per person**
POPULATION **861,000** Density **81 people per sq km** Life Expectancy **45 years** Infant Mortality **132 per 1000**	MAIN PRODUCTS **Groundnuts,** **Fish and fish products** NATIONAL DAY **February 18**
LANGUAGES **English**	ORGANISATIONS **UN, OAU,** **ECOWAS, C**

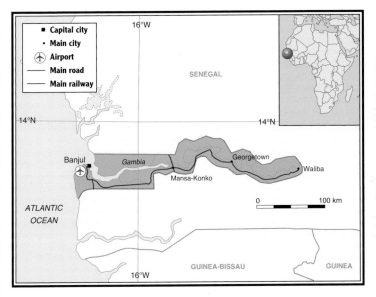

INDEPENDENT from UK 1965; confederation of Senegambia formed with Senegal 1982. Smallest country in Africa; all but its coastline is surrounded by Senegal. Gambia river runs east to west, dividing country. Hot climate with two distinctive seasons: dry, and monsoon rain. Economy currently depends on groundnuts and its by-products, but tourism is developing rapidly as is production of cotton, livestock, fish and rice.

GEORGIA

STATUS **Republic**

AREA **69,700 sq km**

CAPITAL **Tbilisi**

MAIN CITIES **Kutaisi, Rustavi, Batumi**

POPULATION **5,464,000**
Density **78 people per sq km**
Life Expectancy **72 years**
Infant Mortality **20 per 000**

LANGUAGES **Georgian**

RELIGIONS **Orthodox**

CURRENCY **Rouble**

MAIN PRODUCTS **Manganese, Tea, Textiles**

ORGANISATIONS **UN**

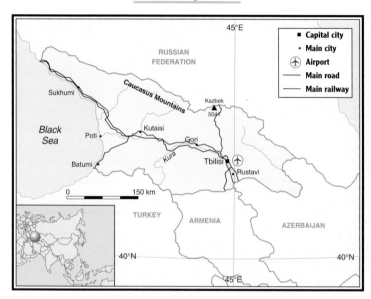

I NDEPENDENT 1991. Mountainous country, much of it forested. Subtropical in the west on the Black Sea; perpetual ice and snow on the crests of the Caucasus Mountains to the north. Rich deposits of coal, oil and manganese, and considerable water-power resources, have encouraged metal and machine-building industries. Agricultural land is scarce and often hard to work, but does produce profitable crops such as tea, grapes, tobacco and citrus fruit. Quest for regional autonomy for ethnic minorities has recently led to violent conflict.

STATUS **Federal Republic**	RELIGIONS **Protestant, Roman Catholic**
AREA **357,868 sq km**	CURRENCY **Mark**
CAPITAL **Berlin**	GNP **20,750 US $ per person**
MAIN CITIES **Hamburg, Munich, Cologne, Frankfurt, Essen**	MAIN PRODUCTS **Machinery, Transport equipment, Office equipment, Chemicals**
POPULATION **79,479,000** Density **222 people per sq km** Life Expectancy **76 years** Infant Mortality **8 per 1000**	NATIONAL DAY **May 23** ORGANISATIONS **UN, EC, OECD, NATO**
LANGUAGES **German**	

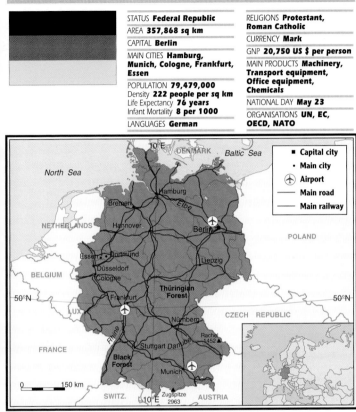

REUNITED in 1990. Most economically powerful member of EC, yet task of reconstruction after reunification is proving more difficult, more expensive and taking longer than expected. Economy was dominated by traditional heavy industry, but services sector is now expanding rapidly. Leading manufacturer and exporter of vehicles, machine tools, electrical and electronic products and consumer goods such as textiles. Also has large stretches of very fertile farmland.

GNP figure is for former West Germany

GHANA

STATUS **Military Regime**
AREA **238,305 sq km**
CAPITAL **Accra**
MAIN CITIES **Sekondi-Takoradi**
POPULATION **15,028,000**
Density **63 people per sq km**
Life Expectancy **56 years**
Infant Mortality **81 per 1000**
LANGUAGES **English**

RELIGIONS **Christian**
CURRENCY **Cedi**
GNP **380 US $ per person**
MAIN PRODUCTS **Cocoa, Gold, Timber**
NATIONAL DAY **March 6**
ORGANISATIONS **UN, OAU, ECOWAS, C**

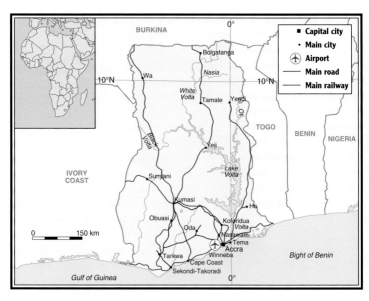

FORMER UK colony of Gold Coast; independent 1957. Landscape varies from tropical rainforest to dry scrubland. Climate hot all year round with moderate to heavy rainfall. Most Ghanaians farm - the principal crop and chief export being cocoa. Other exports include gold and diamonds. A developing industrial base exists around Tema, the largest artificial harbour in Africa. There, local bauxite is smelted into aluminium.

GIBRALTAR see page 216

GREECE

STATUS **Republic**	RELIGIONS **Orthodox**
AREA **131,985 sq km**	CURRENCY **Drachma**
CAPITAL **Athens**	GNP **5,340 US $ per person**
MAIN CITIES **Thessaloniki, Piraiévs, Pátrai**	MAIN PRODUCTS **Clothing, Footwear, Tobacco, Olives and olive oil**
POPULATION **10,123,000** Density **77 people per sq km** Life Expectancy **77 years** Infant Mortality **13 per 1000**	NATIONAL DAY **March 25**
	ORGANISATIONS **UN, EC, NATO, OECD**
LANGUAGES **Greek**	

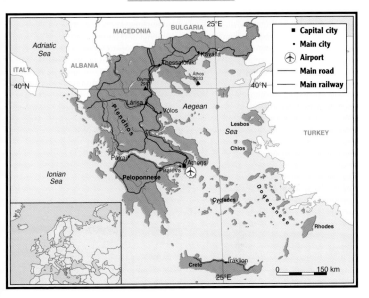

MAINLAND Greece and the many islands are dominated by mountains and sea. Climate is predominantly Mediterranean with hot, dry summers and mild winters. Poor irrigation and drainage mean that much agriculture is localised, but output is now increasing. Main crop, olives, is exported. Surrounding seas provide most of Greece's fish. Very popular tourist destination, which helps the craft industries in textiles, metals and ceramics and other local products.

GREENLAND see page 77

GRENADA

STATUS **Constitutional Monarchy**

AREA **345 sq km**

CAPITAL **St George's**

POPULATION **85,000**
Density **246 people per sq km**
Life Expectancy **69 years**
Infant Mortality **16 per 1000**

LANGUAGES **English**

RELIGIONS **Roman Catholic**

CURRENCY **EC Dollar**

GNP **1,265 US $ per person**

MAIN PRODUCTS **Nutmeg, Bananas, Cocoa**

NATIONAL DAY **February 7**

ORGANISATIONS **UN, OAS, CARICOM, C**

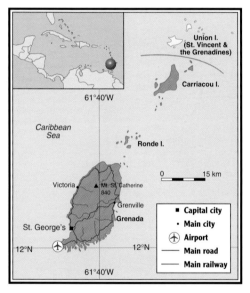

UK COLONY since 1783; independent 1974. Southernmost of the Windward Islands in the east Caribbean Sea. Mountainous and thickly forested, with a settled, warm climate that ensures its tourist industry continues to expand. Main export is bananas; island also famous for its spices, especially nutmeg and cloves. Cocoa is also exported.

GUADELOUPE see page 91
GUAM see page 223

GUATEMALA

STATUS **Republic**	RELIGIONS **Roman Catholic**
AREA **108,890 sq km**	CURRENCY **Quetzal**
CAPITAL **Guatemala City**	GNP **920 US $ per person**
MAIN CITIES **Quezaltenango, Puertos Barrios, Cobán**	MAIN PRODUCTS **Coffee, Sugar, Bananas, Fish, Vegetables**
POPULATION **9,197,000** Density **84 people per sq km** Life Expectancy **65 years** Infant Mortality **48 per 1000**	NATIONAL DAY **September 15**
	ORGANISATIONS **UN, OAS, CACM**
LANGUAGES **Spanish**	

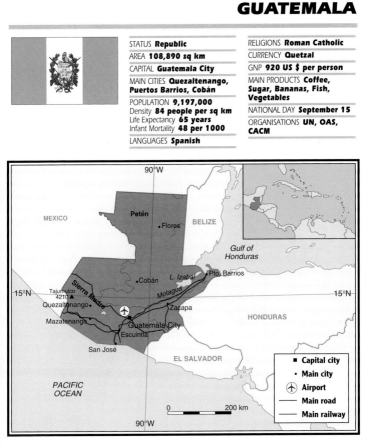

INDEPENDENT from Spain 1821. Borders both the Pacific Ocean and the Caribbean Sea. Mountainous interior covers most of country; has coastal lowlands in the west and east, and a thickly forested area in the north known as the Petén. Agricultural products form bulk of exports, notably coffee, sugar-cane and bananas. Mineral resources including nickel, antimony, lead, silver and, in the north, crude oil are only just beginning to be exploited.

GUERNSEY see page 216

GUINEA

STATUS **Military Regime**	RELIGIONS **Muslim**
AREA **254,855 sq km**	CURRENCY **Franc**
CAPITAL **Conakry**	GNP **430 US $ per person**
MAIN CITIES **Kankan, Labé**	MAIN PRODUCTS **Bauxite, Diamonds, Coffee**
POPULATION **5,756,000** Density **23 people per sq km** Life Expectancy **45 years** Infant Mortality **134 per 1000**	NATIONAL DAY **October 2** ORGANISATIONS **UN, OAU, ECOWAS**
LANGUAGES **French**	

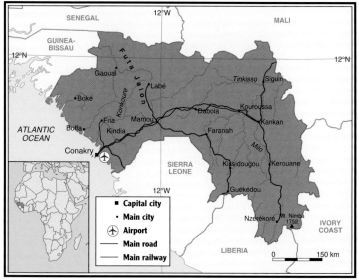

- ■ Capital city
- • Main city
- ✈ Airport
- — Main road
- — Main railway

PART OF former French West Africa; independent 1958. Drowned coastline, lined with mangrove swamps, contrasts strongly with interior highlands containing sources of the Gambia, Niger and Senegal rivers. Agriculture occupies most of the workforce, the main crops being coffee, bananas and pineapple. Some of the largest resources of bauxite (aluminium ore) in the world as well as gold and diamonds. Bauxite and aluminium are exported.

GUINEA-BISSAU

STATUS **Republic**

AREA **36,125 sq km**

CAPITAL **Bissau**

POPULATION **965,000**
Density **27 people per sq km**
Life Expectancy **44 years**
Infant Mortality **140 per 1000**

LANGUAGES **Portuguese**

RELIGIONS **Traditional, Muslim**

CURRENCY **Peso**

GNP **180 US $ per person**

MAIN PRODUCTS **Cashewnuts, Groundnuts**

NATIONAL DAY **September 24**

ORGANISATIONS **UN, OAU, ECOWAS**

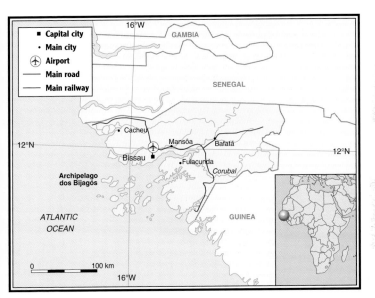

FORMER Portuguese overseas territory; independent 1974. Once a centre for Portuguese slave trade in west Africa. Coast is swampy and lined with mangroves; interior consists of low-lying plain densely covered with rainforest. Climate on coast is hot and humid with high rainfall; interior cooler and drier. Principal crops are groundnuts, cashewnuts and palm oil. Fish, fish products and coconuts also make an important contribution to trade.

GUYANA

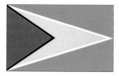

STATUS **Republic**	RELIGIONS **Protestant, Hindu**
AREA **214,970 sq km**	CURRENCY **Dollar**
CAPITAL **Georgetown**	GNP **340 US $ per person**
POPULATION **796,000** Density **4 people per sq km** Life Expectancy **65 years** Infant Mortality **48 per 1000**	MAIN PRODUCTS **Bauxite, Sugar, Shrimps, Gold, Rice**
	NATIONAL DAY **February 23, May 26**
LANGUAGES **English**	ORGANISATIONS **UN, CARICOM, C**

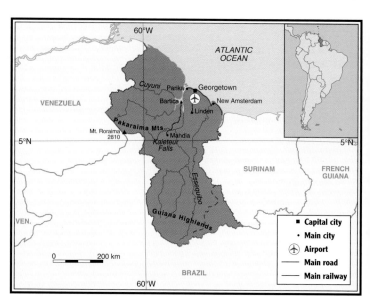

- ■ Capital city
- • Main city
- ✈ Airport
- — Main road
- — Main railway

FORMER UK colony of British Guiana; independent 1966. Coastline on the Atlantic Ocean, the most densely populated area, is flat and marshy, while towards the interior the landscape gradually rises to the Guiana Highlands - a region densely covered in rainforest. Sugar, molasses and rum, once Guyana's main exports, are now less important than bauxite.

STATUS **Republic**

AREA **27,750 sq km**

CAPITAL **Port-au-Prince**

MAIN CITIES **Les Cayes, Jacmel, Jérémie**

POPULATION **6,486,000**
Density **234 people per sq km**
Life Expectancy **57 years**
Infant Mortality **86 per 1000**

LANGUAGES **French, Creole**

RELIGIONS **Roman Catholic**

CURRENCY **Gourde**

GNP **400 US $ per person**

MAIN PRODUCTS **Light manufacturing, Coffee**

NATIONAL DAY **January 1**

ORGANISATIONS **UN, OAS**

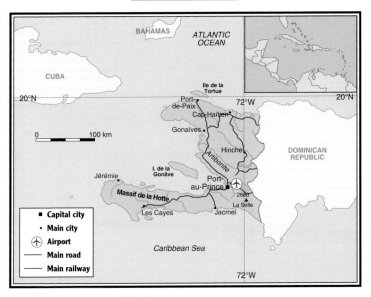

FRENCH colony of Saint Dominique since 1697; independent 1804. Occupies western part of the island of Hispaniola in the Caribbean Sea. Poorest country in Central America. Mountainous with three main ranges. Agriculture restricted to intervening plains. Climate is tropical. Most of workforce are farmers, and coffee is main export. Light manufacturing industries concentrated around the capital.

HONDURAS

STATUS **Republic**	RELIGIONS **Roman Catholic**
AREA **112,085 sq km**	CURRENCY **Lempira**
CAPITAL **Tegucigalpa**	GNP **900 US $ per person**
MAIN CITIES **San Pedro Sula, Choluteca**	MAIN PRODUCTS **Bananas, Coffee, Lead, Zinc, Shellfish**
POPULATION **5,105,000** Density **46 people per sq km** Life Expectancy **66 years** Infant Mortality **57 per 1000**	NATIONAL DAY **September 15**
	ORGANISATIONS **UN, OAS, CACM**
LANGUAGES **Spanish**	

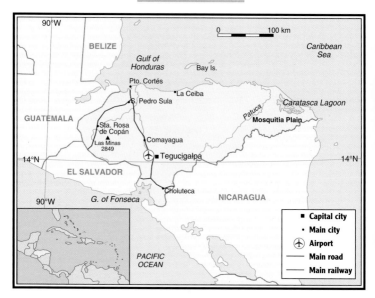

INDEPENDENT from Spain 1821. Poor, sparsely populated country, much of it rugged mountains and high plateaux; in the north-east are low-lying, hot, humid plains on which grow bananas and coffee, accounting for half nation's exports. Other crops include sugar, rice, maize, beans and tobacco. Exploitation of lead, iron, tin and oil may lead to a change in traditional agriculture-based economy. Most industries process local products. Lead, silver and zinc are exported.

HONG KONG see page 219

HUNGARY

STATUS **Republic**

AREA **93,030 sq km**

CAPITAL **Budapest**

MAIN CITIES **Debrecen, Miskolc, Pécs, Szeged, Gyor**

POPULATION **10,553,000**
Density **113 people per sq km**
Life Expectancy **72 years**
Infant Mortality **17 per 1000**

LANGUAGES **Hungarian**

RELIGIONS **Roman Catholic**

CURRENCY **Forint**

GNP **2,560 US $ per person**

MAIN PRODUCTS **Semifinished products, Raw and basic materials**

NATIONAL DAY **March 15, August 20, October 23**

ORGANISATIONS **UN**

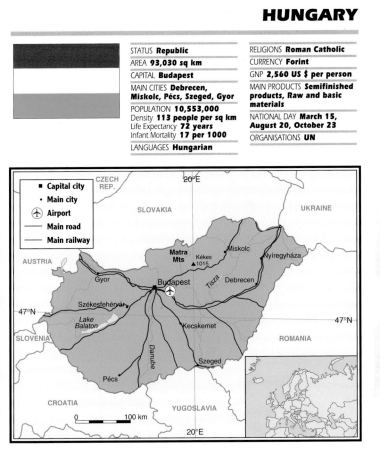

REPUBLIC 1918. Undulating, fertile plains, which are bisected by River Danube, can become very hot in summer; winters are severe. Only substantial mineral resource is bauxite; agriculture now relatively unimportant. Massive drive for industrialisation has fundamentally transformed structure of economy since 1945. Capital and consumer industries were developed and during the 1980s engineering accounted for more than half total industrial output. Despite fall of communism in 1990 economic and political outlook is unsettled.

ICELAND

STATUS **Republic**	RELIGIONS **Protestant**
AREA **102,820 sq km**	CURRENCY **Króna**
CAPITAL **Reykjavik**	GNP **21,240 US $ per person**
POPULATION **255,000**	MAIN PRODUCTS **Fish and fish**
Density **2 people per sq km**	**products, Aluminium,**
Life Expectancy **78 years**	**Ferrosilicon**
Infant Mortality **5 per 1000**	NATIONAL DAY **June 17**
LANGUAGES **Icelandic**	ORGANISATIONS **UN, EFTA, OECD, NATO**

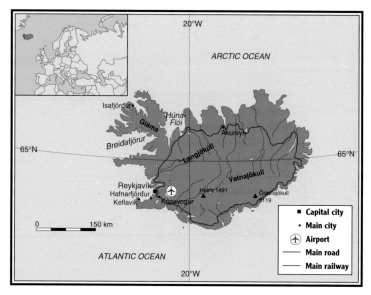

- ■ Capital city
- • Main city
- ✈ Airport
- — Main road
- — Main railway

HOME-RULE from Denmark 1918; independent 1944. Northernmost island in Europe. Landscape is entirely volcanic - compacted volcanic ash has been eroded by wind. Substantial ice sheets, lava fields, still-active volcanoes, geysers and hot springs. Climate is cold and vegetation sparse. Main industry is fishing, and almost all exports consist of fish and fish products.

INDIA

STATUS **Federal Republic**

AREA **3,166,830 sq km**

CAPITAL **New Delhi**

MAIN CITIES **Calcutta, Bombay, Madras, Bangalore, Hyderabad**

POPULATION **827,057,000**
Density **261 people per sq km**
Life Expectancy **60 years**
Infant Mortality **88 per 1000**

LANGUAGES **Hindi, English**

RELIGIONS **Hindu**

CURRENCY **Rupee**

GNP **350 US $ per person**

MAIN PRODUCTS **Precious stones, Machinery, Clothing, Transport equipment**

NATIONAL DAY **January 26**

ORGANISATIONS **UN, CP, C**

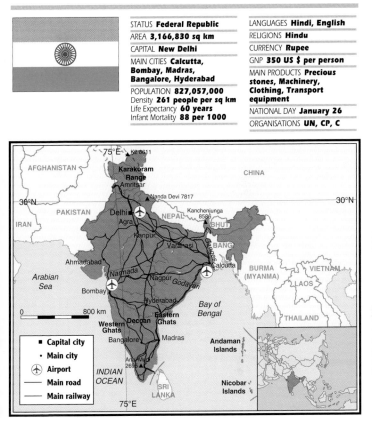

INDEPENDENT from UK 1947; Goa joined 1961 and Sikkim 1975. World's second largest population. In the north are the Himalayas, the world's highest mountain range, lush vegetation, abundant water supply and hot climate. To the south lie one of the world's most fertile regions, although it is liable to flooding, and the Deccan Plateau. Rice, wheat, cotton, jute, tobacco, tea and sugar are main crops. Most people are subsistence farmers. Immense reserves of timber, coal, iron ore, nickel and off-shore oil. Tourism also important.

INDONESIA

STATUS **Republic**

AREA **1,919,445 sq km**

CAPITAL **Jakarta**

MAIN CITIES **Surabaya, Bandung, Medan, Semerang, Ujung Pandang**

POPULATION **179,300,000**
Density **93 people per sq km**
Life Expectancy **63 years**
Infant Mortality **65 per 1000**

LANGUAGES **Bahasa Indonesia**

RELIGIONS **Muslim**

CURRENCY **Rupiah**

GNP **490 US $ per person**

MAIN PRODUCTS **Crude oil, Natural gas, Timber, Clothing, Rubber**

NATIONAL DAY **August 17**

ORGANISATIONS **UN, ASEAN, OPEC, CP**

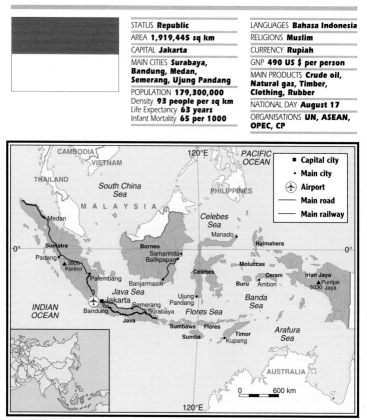

FORMER Dutch East Indies; independent 1949; Western New Guinea joined 1963 (now called Irian Jaya). Comprises thousands of islands, some uninhabited, along the equator. Climate is tropical: hot, wet and subject to monsoons. Fourth largest population in the world. Most people live on Java and farm along the coasts or in river valleys, but crops produced are hardly enough for rising population. Fishing industry needs developing. Timber, oil and tourism earn important foreign revenue. Rich mineral deposits not yet fully exploited.

IRAN

STATUS **Islamic Republic**	LANGUAGES **Persian**
AREA **1,648,000 sq km**	RELIGIONS **Shi'a Muslim**
CAPITAL **Tehran**	CURRENCY **Rial**
MAIN CITIES **Mashhad, Isfahan, Tabriz, Shiraz, Abadan**	GNP **1,800 US $ per person**
	MAIN PRODUCTS **Petroleum products, Carpets, Hides, Fruit**
POPULATION **54,608,000** Density **33 people per sq km** Life Expectancy **67 years** Infant Mortality **40 per 1000**	NATIONAL DAY **February 11**
	ORGANISATIONS **UN, OPEC, CP**

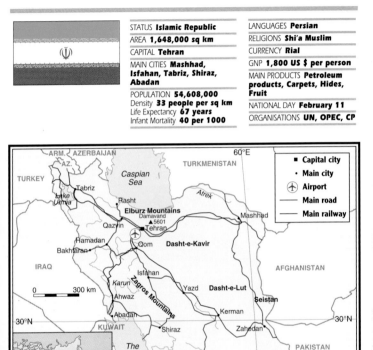

LARGE mountainous country situated between the Caspian Sea and The Gulf. Climate ranges from very cold to extremely hot and rainfall from high to almost zero. Rich in oil and gas; other substantial mineral deposits relatively underdeveloped. Agricultural conditions are poor except around Caspian Sea. Wheat is main crop, though fruit (especially dates) and nuts are grown and exported. Main livestock is sheep and goats. Economic growth, especially in Gulf oil industry, seriously restricted in 1980s during war with Iraq.

IRAQ

STATUS **Republic**

AREA **438,445 sq km**

CAPITAL **Baghdād**

MAIN CITIES **Basra, Mosul, Sulaymañiyah**

POPULATION **18,920,000**
Density **43 people per sq km**
Life Expectancy **66 years**
Infant Mortality **56 per 1000**

LANGUAGES **Arabic, Kurdish**

RELIGIONS **Muslim**

CURRENCY **Dinar**

GNP **1,940 US $ per person**

MAIN PRODUCTS **Crude oil**

NATIONAL DAY **July 17**

ORGANISATIONS **UN, OPEC, AL**

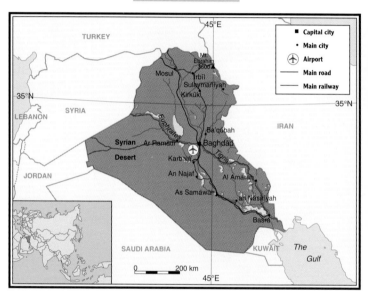

MOSTLY marsh and mountain, yet there are large areas of fertile land between the two great rivers of the Tigris and Euphrates. Climate is mainly arid with low, unreliable rainfall. Summers are very hot; winters cold. Oil is main export; the major petro-chemical complexes are around the oilfields. Light industry is situated around the capital. Wars in 1980s and 1990s have placed great strains on economy with exports of oil and gas severely restricted. Iraq will take some time to recover from the damage caused to its infrastructure.

STATUS **Republic**	RELIGIONS **Roman Catholic**
AREA **68,895 sq km**	CURRENCY **Punt**
CAPITAL **Dublin**	GNP **8,500 US $ per person**
MAIN CITIES **Cork, Limerick, Galway**	MAIN PRODUCTS **Machinery, Transport equipment, Food, Chemical products**
POPULATION **3,503,000** Density **51 people per sq km** Life Expectancy **75 years** Infant Mortality **8 per 1000**	NATIONAL DAY **March 17**
	ORGANISATIONS **UN, EC, OECD**
LANGUAGES **English, Irish**	

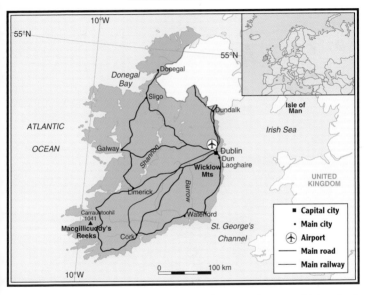

INDEPENDENT 1921. Occupies all but north-east of island of Ireland. Cool, damp climate creates rich pastureland. Livestock farming predominates. Meat and dairy produce processed in small market towns, where there are also breweries and mills. Large-scale manufacturing centred around Dublin, the capital and main port. Reserves of oil, natural gas and peat and deposits of lead and zinc.

ISLE OF MAN see page 216

ISRAEL

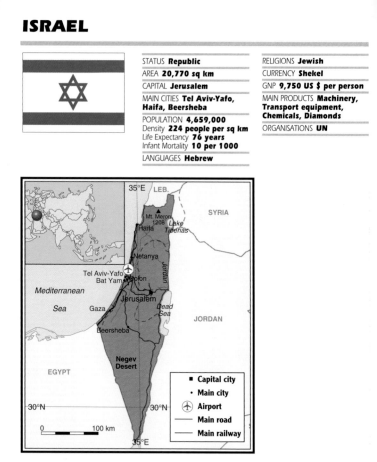

STATUS **Republic**	RELIGIONS **Jewish**
AREA **20,770 sq km**	CURRENCY **Shekel**
CAPITAL **Jerusalem**	GNP **9,750 US $ per person**
MAIN CITIES **Tel Aviv-Yafo, Haifa, Beersheba**	MAIN PRODUCTS **Machinery, Transport equipment, Chemicals, Diamonds**
POPULATION **4,659,000** Density **224 people per sq km** Life Expectancy **76 years** Infant Mortality **10 per 1000**	ORGANISATIONS **UN**
LANGUAGES **Hebrew**	

FORMERLY under UK administration; independent 1948. Coastal plain in the west; Galilee Highlands in the north and Negev Desert in the south. Economic development is most advanced in Middle East. Manufacturing, particularly diamond finishing and electronics, and mining are most important industries. Flourishing agricultural industry exports fruit, flowers and vegetables to western Europe.

STATUS **Republic**	LANGUAGES **Italian**
AREA **301,245 sq km**	RELIGIONS **Roman Catholic**
CAPITAL **Rome**	CURRENCY **Lira**
MAIN CITIES **Milan, Naples, Turin, Genoa, Palermo, Bologna**	GNP **15,150 US $ per person**
POPULATION **57,662,000** Density **191 people per sq km** Life Expectancy **76 years** Infant Mortality **9 per 1000**	MAIN PRODUCTS **Machinery, Transport equipment, Chemicals, Clothing**
	NATIONAL DAY **June 2**
	ORGANISATIONS **UN, EC, OECD, NATO**

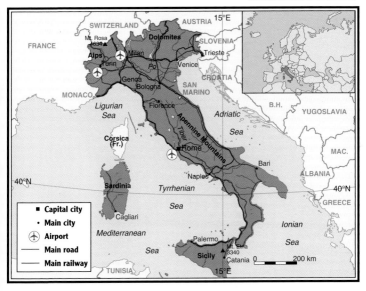

MAINLY mountainous or hilly landscape with the north dominated by the flat plain of the River Po climbing to the high Alps. Climate varies from hot summers and mild winters in the south and lowland areas, to mild summers and cold winters in the Alps. Agriculture thrives with cereals, vegetables, olives and wines the principal crops. Few natural resources. Manufacturing of cars, machine tools, textile machinery and engineering, mainly in the north, is expanding rapidly. Rising imbalance in income and investment between north and south.

IVORY COAST

STATUS **Republic**
AREA **322,465 sq km**
CAPITAL **Yamoussoukro**
MAIN CITIES **Abidjan, Bouaké, Daloa**
POPULATION **11,998,000**
Density **37 people per sq km**
Life Expectancy **54 years**
Infant Mortality **87 per 1000**
LANGUAGES **French**

RELIGIONS **Muslim, Roman Catholic**
CURRENCY **CFA Franc**
GNP **790 US $ per person**
MAIN PRODUCTS **Cocoa, Coffee, Cotton, Timber**
NATIONAL DAY **December 7**
ORGANISATIONS **UN, OAU, ECOWAS**

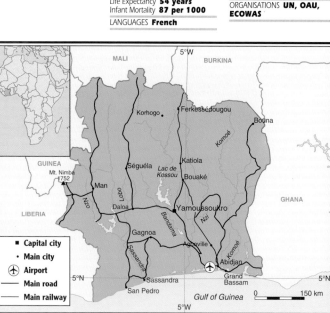

PART OF former French West Africa; independent 1960. Stretches from low plains in the south to plateaux in the north. Climate is tropical with rainfall all year round in the south. Most people engage in agriculture, producing rice, cassava, maize, sorghum, plantains and yams. Exports include coffee, timber and cocoa. Main industrial area and leading port is centred on Abidjan. Important industries are food processing, textiles and timber products.

STATUS **Constitutional Monarchy**

AREA **11,425 sq km**

CAPITAL **Kingston**

MAIN CITIES **Spanish Town, Montego Bay**

POPULATION **2,420,000**
Density **212 people per sq km**
Life Expectancy **74 years**
Infant Mortality **14 per 1000**

LANGUAGES **English**

RELIGIONS **Protestant**

CURRENCY **Dollar**

GNP **1,260 US $ per person**

MAIN PRODUCTS **Bauxite, Sugar, Bananas**

NATIONAL DAY **1st Monday in August**

ORGANISATIONS **UN, OAS, CARICOM, C**

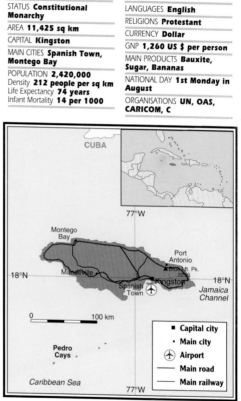

UK COLONY since 1670; independent 1962. Situated in the centre of the Greater Antilles chain of islands in the Caribbean Sea. Formed from peaks of a submerged mountain range. Climate is tropical with very heavy rainfall. Plentiful supply of tropical fruits such as melons, bananas and guavas. Principal crops include sugar-cane, bananas and coffee. Rich in bauxite, which provides over half foreign-exchange earnings. Main manufacturing industries are food processing, textiles, cement and agricultural machinery.

JAPAN

STATUS **Constitutional Monarchy**	LANGUAGES **Japanese**
AREA **369,700 sq km**	RELIGIONS **Shintoist, Buddhist**
CAPITAL **Tokyo**	CURRENCY **Yen**
MAIN CITIES **Yokohama, Osaka, Nagoya, Sapporo, Kyoto, Kobe**	GNP **23,730 US $ per person**
	MAIN PRODUCTS **Motor vehicles, Office machinery, Iron and steel products, Scientific equipment**
POPULATION **123,537,000** Density **334 people per sq km** Life Expectancy **79 years** Infant Mortality **5 per 1000**	NATIONAL DAY **December 23**
	ORGANISATIONS **UN, OECD, CP**

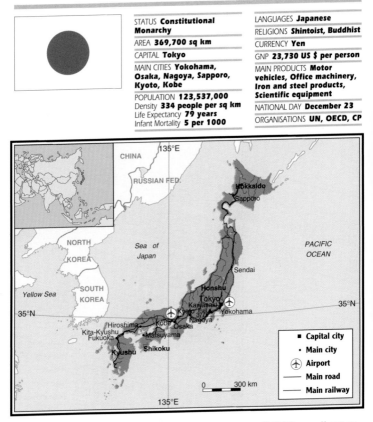

CONSISTS of four main islands and some 3,000 small ones. Mountainous, heavily forested land with small, fertile patches on which rice is grown, often by part-time farmers, to feed whole population. Climate ranges from harsh to tropical. Subject to monsoons, earthquakes, typhoons and tidal waves. Fishing fleet largest in the world. Leading economic power. Most people are involved in industry. Principal exports are electronic, electrical and optical equipment. Production of nuclear power, coal, oil and natural gas is being increased.

JERSEY see page 216

JORDAN

STATUS **Constitutional Monarchy**

AREA **96,000 sq km**

CAPITAL **Amman**

MAIN CITIES **Irbid, Zarqa**

POPULATION **4,010,000**
Density **42 people per sq km**
Life Expectancy **68 years**
Infant Mortality **36 per 1000**

LANGUAGES **Arabic**

RELIGIONS **Sunni Muslim**

CURRENCY **Dinar**

GNP **1,730 US $ per person**

MAIN PRODUCTS **Chemicals, Phosphates, Basic manufactured goods**

NATIONAL DAY **May 25**

ORGANISATIONS **UN, AL**

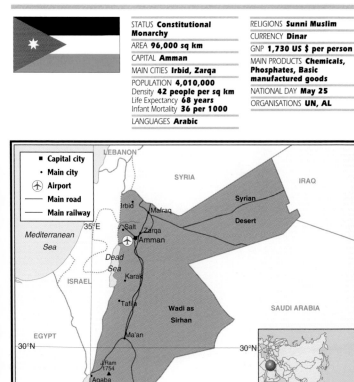

FORMERLY under British mandate; independent 1946. One of the remaining kingdoms in the Middle East. Mostly desert, with fertile pockets. Temperatures are extremely high in the valleys, but cooler and wetter in the west. Amman is the manufacturing centre, processing bromide and potash from the Dead Sea. Other important industries are food processing and textiles.

KAZAKHSTAN

STATUS **Republic**

AREA **2,717,300 sq km**

CAPITAL **Alma-Ata**

MAIN CITIES **Karaganda, Semipalatinsk, Chimkent**

POPULATION **16,742,000**
Density **6 people per sq km**
Life Expectancy **69 years**
Infant Mortality **26 per 1000**

LANGUAGES **Kazakh**

RELIGIONS **Sunni Muslim**

CURRENCY **Rouble**

MAIN PRODUCTS **Grain, Minerals, Clothing**

ORGANISATIONS **UN**

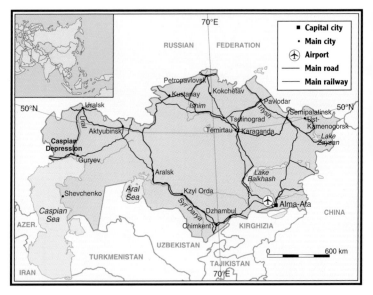

INDEPENDENT 1991. Consists of lowlands, hilly plains and plateaux, with a small, mountainous area. Continental climate with hot summers and very cold winters. Exceptionally rich in raw materials. Rapid industrialisation in recent years has focused on iron and steel, cement, chemicals, fertilizers and consumer goods. Most agricultural land used for pasture. Economic prospects appear favourable yet environmental problems caused for example by Soviet exploitation of the Aral Sea for irrigation have still to be tackled.

KENYA

STATUS **Republic**

AREA **582,645 sq km**

CAPITAL **Nairobi**

MAIN CITIES **Mombasa, Kisumu, Nakuru, Malindi**

POPULATION **24,032,000**
Density **41 people per sq km**
Life Expectancy **61 years**
Infant Mortality **64 per 1000**

LANGUAGES **Swahili, English**

RELIGIONS **Roman Catholic, Protestant**

CURRENCY **Shilling**

GNP **380 US $ per person**

MAIN PRODUCTS **Tea, Coffee, Horticulture, Petroleum products**

NATIONAL DAY **December 12**

ORGANISATIONS **UN, OAU, PTA, C**

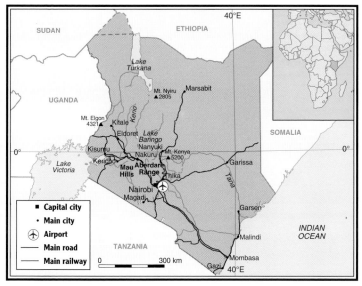

INDEPENDENT from UK 1963. Lies on the equator but as most of Kenya is on a high plateau the temperatures range from moderate to high. Rainfall is low to high depending on altitude. Poor soil and a dry climate mean that little land is cultivated. Nonetheless exports are dominated by farm produce: coffee, tea, sisal and meat. Nairobi and Mombasa are main manufacturing centres. Tourist industry is growing. Electricity is generated from geothermal sources and hydro-electric power stations on the Tana River.

KIRGHIZIA

STATUS **Republic**

AREA **198,500 sq km**

CAPITAL **Bishkek**

MAIN CITIES **Osh, Naryn**

POPULATION **4,394,000**
Density **22 people per sq km**
Life Expectancy **68 years**
Infant Mortality **33 per 1000**

LANGUAGES **Kirgiz**

RELIGIONS **Sunni Muslim**

CURRENCY **Som**

MAIN PRODUCTS **Livestock, Machinery**

ORGANISATIONS **UN**

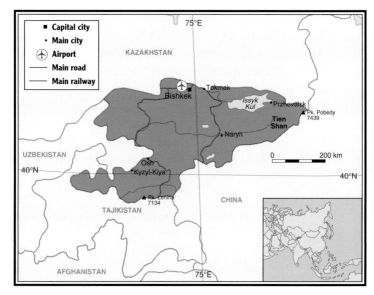

Independent 1991. This mountainous country is located in the heart of Asia. Economy was dominated by agriculture, especially livestock farming, but this declined as Kirghizia underwent rapid industrialisation in the 20th century. Is now a major producer of machinery and hydro-electric power. Coal, antimony and mercury are mined. Cultivation of cotton, sugar beet, tobacco and opium poppies is expanding and providing basis for growing processing industry.

STATUS **Republic**	RELIGIONS **Protestant, Roman Catholic**
AREA **684 sq km**	CURRENCY **Australian Dollar**
CAPITAL **Bairiki on Tarawa Atoll**	GNP **650 US $ per person**
POPULATION **66,000**	MAIN PRODUCTS **Copra**
Density **96 people per sq km**	NATIONAL DAY **July 12**
Life Expectancy **53 years**	
Infant Mortality **110 per 1000**	ORGANISATIONS **SPF, C**
LANGUAGES **English**	

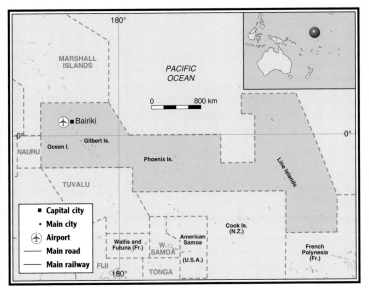

Part of the former Gilbert and Ellice Islands Colony; separation 1975; independent 1979. Comprises 16 Gilbert Islands, eight Phoenix Islands, three Line Islands and Ocean Island, spread over central and west Pacific Ocean. Temperature is consistently high. Principal crops are coconut, breadfruit, bananas and babai (a coarse vegetable). Copra is only major export. Main imports are machinery and manufactured goods.

KUWAIT

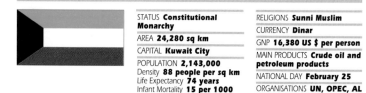

STATUS **Constitutional Monarchy**

AREA **24,280 sq km**

CAPITAL **Kuwait City**

POPULATION **2,143,000**
Density **88 people per sq km**
Life Expectancy **74 years**
Infant Mortality **15 per 1000**

LANGUAGES **Arabic**

RELIGIONS **Sunni Muslim**

CURRENCY **Dinar**

GNP **16,380 US $ per person**

MAIN PRODUCTS **Crude oil and petroleum products**

NATIONAL DAY **February 25**

ORGANISATIONS **UN, OPEC, AL**

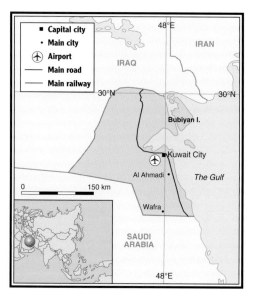

- ■ **Capital city**
- • **Main city**
- ✈ **Airport**
- — **Main road**
- — **Main railway**

48°E

IRAN

IRAQ

30°N

Bubiyan I.

✈ ■Kuwait City

Al Ahmadi •

The Gulf

0   150 km

Wafra •

SAUDI ARABIA

48°E

UK **PROTECTORATE** since 1899; independent 1961. Comprises low, undulating desert, with very high summer temperatures and minimal rainfall. Vegetables for domestic consumption are grown in expanding irrigated areas. Since discovery of oil Kuwait has been transformed into one of the world's wealthiest nations. Other industries include natural gas, fishing (particularly shrimp), food processing, chemicals and building materials. Recent attempted annexation of Kuwait by Iraq has had severe effects on Kuwait's economy.

LAOS

STATUS **Republic**
AREA **236,725 sq km**
CAPITAL **Vientiane**
POPULATION **4,139,000**
Density **17 people per sq km**
Life Expectancy **51 years**
Infant Mortality **97 per 1000**
LANGUAGES **Lao**

RELIGIONS **Buddhist**
CURRENCY **Kip**
GNP **170 US $ per person**
MAIN PRODUCTS **Energy, Timber, Coffee, Tin**
NATIONAL DAY **December 2**
ORGANISATIONS **UN, CP**

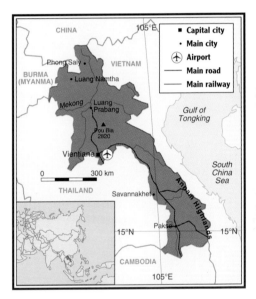

FORMER French protectorate; independent 1949. This poor, landlocked country in Indo-China has moderate to high temperatures. Most of the sparse population are farmers growing small amounts of rice, maize, sweet potatoes and tobacco. Major exports are tin and timber, the latter floated down the Mekong River. Almost constant warfare for 50 years has hindered any possible industrial development.

LATVIA

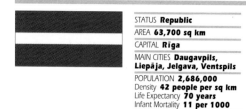

STATUS **Republic**

AREA **63,700 sq km**

CAPITAL **Rīga**

MAIN CITIES **Daugavpils, Liepāja, Jelgava, Ventspils**

POPULATION **2,686,000**
Density **42 people per sq km**
Life Expectancy **70 years**
Infant Mortality **11 per 1000**

LANGUAGES **Latvian**

RELIGIONS **Protestant**

CURRENCY **Latvian Rouble**

MAIN PRODUCTS **Machinery, Clothing, Paper**

ORGANISATIONS **UN**

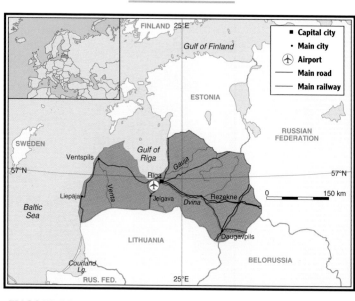

INCORPORATED into USSR 1940; independent 1991. Situated on the shores of the Baltic Sea and the Gulf of Riga. Forests, meadows, swamps and wasteland predominate. Farmland supports dairy and meat production and grain crops. Climate is oceanic: windy, cloudy and humid. Minimal mineral resources. Machine-building and metal engineering of for example refrigerators, motorcycles and ships are chief manufacturing activities. Fundamental reforms are needed owing to environmental damage, commercial unprofitability and collapse of communism.

STATUS **Republic**		RELIGIONS **Muslim, Christian**
AREA **10,400 sq km**		CURRENCY **Pound**
CAPITAL **Beirut**		GNP **690 US $ per person**
MAIN CITIES **Tripoli, Zahle, Saida**		MAIN PRODUCTS **Jewellery, Clothing, Pharmaceuticals, Metal products**
POPULATION **2,701,000**		
Density **260 people per sq km**		NATIONAL DAY **November 22**
Life Expectancy **67 years**		ORGANISATIONS **UN, AL**
Infant Mortality **40 per 1000**		
LANGUAGES **Arabic**		

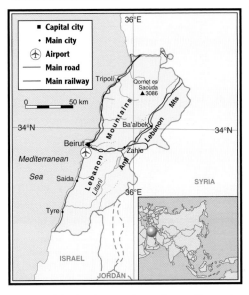

INDEPENDENT state 1920; independent from French administration 1941. Divides into narrow, coastal plain, and narrow, fertile, interior plateau between central Lebanon Mountains and Anti-Lebanon Mountains in the east. Mediterranean climate with moderate rainfall on coast, high in the mountains. Trade and tourism have been severely affected by civil war since 1970s. Agriculture accounts for nearly half all employed people. Cement, fertilisers, jewellery, sugar and tobacco products are all manufactured on a small scale.

LESOTHO

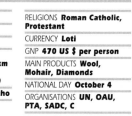

STATUS **Military Regime**	RELIGIONS **Roman Catholic, Protestant**
AREA **30,345 sq km**	CURRENCY **Loti**
CAPITAL **Maseru**	GNP **470 US $ per person**
POPULATION **1,774,000** Density **58 people per sq km** Life Expectancy **59 years** Infant Mortality **89 per 1000**	MAIN PRODUCTS **Wool, Mohair, Diamonds**
	NATIONAL DAY **October 4**
LANGUAGES **English, Sesotho**	ORGANISATIONS **UN, OAU, PTA, SADC, C**

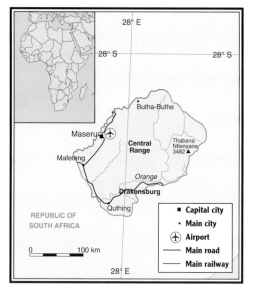

UK PROTECTORATE of Basutoland since 1884; independent 1966. This rugged, mountainous country is completely encircled by South Africa. In the east is southern Africa's highest mountain, Thabana Ntlenyana. Because of the terrain, agriculture is limited to the lowlands and foothills. Sorghum, wheat, barley, maize, oats and legumes are main crops. Cattle, sheep and goats graze on the highlands.

LIBERIA

STATUS **Republic**
AREA **111,370 sq km**
CAPITAL **Monrovia**
POPULATION **2,607,000**
Density **23 people per sq km**
Life Expectancy **55 years**
Infant Mortality **126 per 1000**
LANGUAGES **English**

RELIGIONS **Christian**
CURRENCY **Dollar**
GNP **450 US $ per person**
MAIN PRODUCTS **Iron ore,
Rubber, Timber, Diamonds,
Coffee**
NATIONAL DAY **July 26**
ORGANISATIONS **UN, OAU,
ECOWAS**

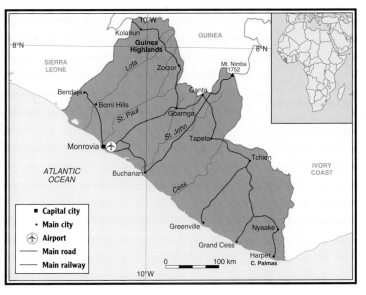

FOUNDED 1822 by American Colonisation Society as settlement for freed American slaves; independent 1847. Only nation in Africa never to have been ruled by a foreign power. Hot, humid coastal plain, with savannah vegetation and mangrove swamps, rises gently towards the Guinea Highlands. Interior is densely covered by tropical rainforest. Rubber, formerly Liberia's main export, has now been supplemented by iron ore. World's largest fleet of merchant ships owing to its flag of convenience tax regime.

LIBYA

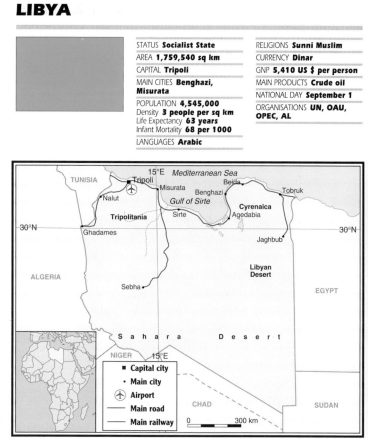

STATUS **Socialist State**	RELIGIONS **Sunni Muslim**
AREA **1,759,540 sq km**	CURRENCY **Dinar**
CAPITAL **Tripoli**	GNP **5,410 US $ per person**
MAIN CITIES **Benghazi, Misurata**	MAIN PRODUCTS **Crude oil**
POPULATION **4,545,000**	NATIONAL DAY **September 1**
Density **3 people per sq km**	ORGANISATIONS **UN, OAU, OPEC, AL**
Life Expectancy **63 years**	
Infant Mortality **68 per 1000**	
LANGUAGES **Arabic**	

INDEPENDENT from Italy 1951. Situated on the lowlands of north Africa and northern Sahara Desert. Almost totally covered in hot, dry desert, with minimal rainfall. Coastal plains with their more moist Mediterranean climate are most densely populated and main cultivated areas. Here are grown grapes, groundnuts, oranges, wheat and barley. Dates are cultivated in the desert oases. Exploitation of oil since 1960s has transformed economy - oil now accounting for almost all exports.

LIECHTENSTEIN

STATUS **Constitutional Monarchy**

AREA **160 sq km**

CAPITAL **Vaduz**

POPULATION **29,000**
Density **181 people per sq km**
Life Expectancy **70 years**
Infant Mortality **3 per 1000**

LANGUAGES **German**

RELIGIONS **Roman Catholic**

CURRENCY **Swiss Franc**

GNP **21,000 US $ per person**

MAIN PRODUCTS **Machinery, Metal products**

NATIONAL DAY **August 15**

ORGANISATIONS **UN, EFTA**

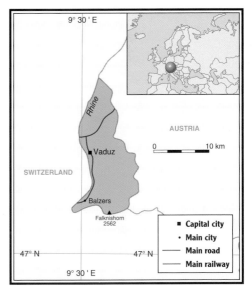

SITUATED in central Alps between Switzerland and Austria. One of the smallest states in Europe. Divides into plains around the Rhine to the north and Alpine mountain ranges to the south where cattle are reared. Other main sources of revenue comprise light industry chiefly manufacture of precision instruments, also textile production, food products and tourism.

LITHUANIA

STATUS **Republic**	LANGUAGES **Lithuanian**
AREA **65,200 sq km**	RELIGIONS **Roman Catholic**
CAPITAL **Vilnius**	CURRENCY **Litas**
MAIN CITIES **Kaunas, Klaipéda, Šiauliai, Panevėžys**	MAIN PRODUCTS **Crude oil, Food products, Consumer electronics**
POPULATION **3,731,000** Density **57 people per sq km** Life Expectancy **71 years** Infant Mortality **11 per 1000**	ORGANISATIONS **UN**

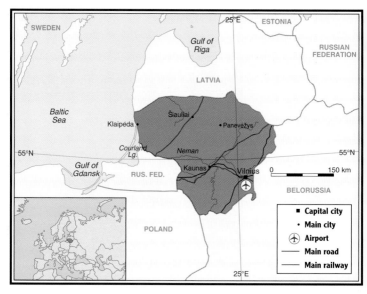

INCORPORATED into USSR 1940; independent 1991. Low-lying plain on the shores of the Baltic Sea. Climate is transitional between oceanic type of western Europe and continental conditions prevailing farther east. Massive drive for industrialisation during the Soviet period has done enormous damage to the environment and has failed to create competitive enterprises that can survive under market conditions. Production has fallen and unemployment risen. Agriculture, with its emphasis on meat and dairy products, still awaits decollectivisation.

LUXEMBOURG

STATUS **Constitutional Monarchy**

AREA **2,585 sq km**

CAPITAL **Luxembourg**

POPULATION **381,000**
Density **147 people per sq km**
Life Expectancy **75 years**
Infant Mortality **9 per 1000**

LANGUAGES **Luxembourgian, French, German**

RELIGIONS **Roman Catholic**

CURRENCY **Franc**

GNP **24,860 US $ per person**

MAIN PRODUCTS **Metal products, Machinery, Plastics, Textiles**

NATIONAL DAY **June 23**

ORGANISATIONS **UN, EC, OECD, NATO**

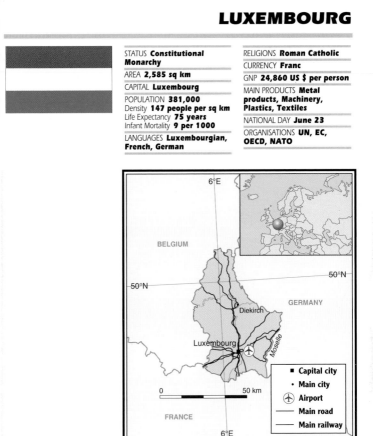

INDEPENDENT 1815. In the north is the Oesling region - an extension of the high Ardennes plateau - which is cut through by thickly forested river valleys. Gutland in the south has rolling, lush pastureland. Climate is mild and temperate. Just over half the land is arable, mainly cereals, dairy produce, potatoes and grapes (for wine). Iron ore reserves in the south. Major industries are steel, textiles, chemicals, metal goods and pharmaceutical products.

MACAU see page 171

MACEDONIA

STATUS **Republic**
AREA **25,713 sq km**
CAPITAL **Skopje**
MAIN CITIES **Ohrid, Prilep**
POPULATION **2,034,000**
Density **79 people per sq km**
Life Expectancy **72 years**
Infant Mortality **35 per 1000**
LANGUAGES **Macedonian**

RELIGIONS **Orthodox**
CURRENCY **Dinar**
MAIN PRODUCTS **Agricultural products**

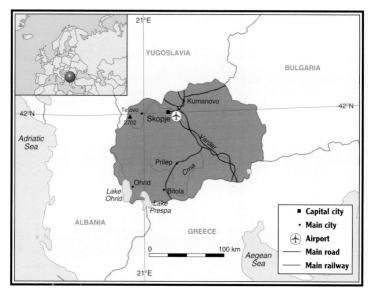

THOUGH not universally recognised as a separate country, de facto separated from Yugoslavia 1992. Landlocked mountainous country divided by two major river valleys. There are dry summers but heavy rainfall in the mountains. One of the least developed republics of the former Yugoslavia. Rich in agricultural land but industrial development has suffered from setbacks. Has large Albanian and Gypsy minorities. Declaration of independence has caused concern in Greece and Bulgaria who fear more ethnic conflict in the Balkans.

MADAGASCAR

STATUS **Republic**

AREA **594,180 sq km**

CAPITAL **Antananarivo**

MAIN CITIES **Toamasina, Fianarantsoa, Mahajanga**

POPULATION **11,197,000**
Density **19 people per sq km**
Life Expectancy **56 years**
Infant Mortality **120 per 1000**

LANGUAGES **Malagasy, French**

RELIGIONS **Christian, Traditional**

CURRENCY **Franc**

GNP **230 US $ per person**

MAIN PRODUCTS **Coffee, Vanilla, Cloves, Sugar**

NATIONAL DAY **June 26**

ORGANISATIONS **UN, OAU**

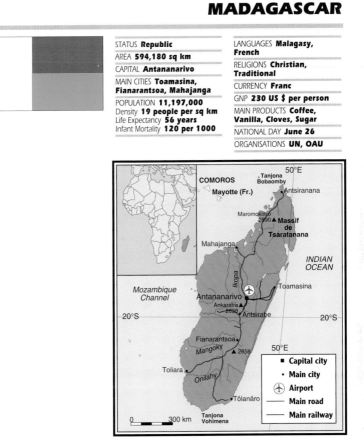

FRENCH colony since 1896; independent 1960. World's fourth largest island; off East African coast. High, central plateau of savannah and arid scrub; desert in the south. Much of hot, humid, east coast is covered in tropical rainforest. Most people farm yet very little land is cultivated. Coffee, rice and cassava are main products. Habitats of unique plant and animal life are threatened by rapid development of forestry and soil erosion.

MADEIRA see page 171

MALAWI

STATUS **Republic**	RELIGIONS **Christian**
AREA **94,080 sq km**	CURRENCY **Kwacha**
CAPITAL **Lilongwe**	GNP **180 US $ per person**
MAIN CITIES **Blantyre**	MAIN PRODUCTS **Tobacco, Tea, Sugar, Groundnuts**
POPULATION **8,289,000**	
Density **88 people per sq km**	NATIONAL DAY **July 6**
Life Expectancy **49 years**	ORGANISATIONS **UN, OAU,**
Infant Mortality **138 per 1000**	**PTA, SADC, C**
LANGUAGES **English, Chichewa**	

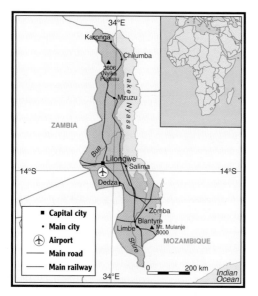

UK PROTECTORATE since 1891; named Nyasaland 1907; Federation of Rhodesia and Nyasaland 1953; independent 1964. Area around Lake Nyasa is hot and humid with swampy vegetation, highlands to the west and south-east have cooler conditions. Most people are farmers. Maize is chief subsistence crop; tea, tobacco, sugar and groundnuts cash crops. Underexploited deposits of coal and bauxite. Manufacturing industry, fuelled by hydro-electric power, produces consumer goods and building and construction materials.

STATUS **Federal Constitutional Monarchy**	LANGUAGES **Bahasa Malay**
AREA **332,965 sq km**	RELIGIONS **Muslim**
CAPITAL **Kuala Lumpur**	CURRENCY **Ringgit**
MAIN CITIES **Ipoh, George Town, Jahor Baharu, Kota Baharu, Kuala Terengganu**	GNP **2,130 US $ per person**
	MAIN PRODUCTS **Electronic machinery, Crude oil, Timber, Rubber**
POPULATION **17,861,000** Density **54 people per sq km** Life Expectancy **71 years** Infant Mortality **20 per 1000**	NATIONAL DAY **August 31**
	ORGANISATIONS **UN, ASEAN, CP, C**

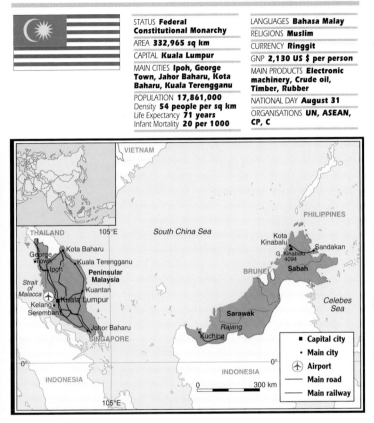

INDEPENDENT from UK 1957. Consists of Peninsular Malaysia and Sabah and Sarawak on island of Borneo. Mountainous landscape covered with lush tropical rainforest. Climate is hot and humid all year round. One of world's main tin and rubber producers; also leading source of palm oil, bauxite and gold. Most manufacturers process local materials. Pigs, cattle, goats, buffaloes and sheep are also important locally. Export crops include pineapples, tobacco, cocoa and spices.

MALDIVES

STATUS **Republic**	RELIGIONS **Sunni Muslim**
AREA **298 sq km**	CURRENCY **Rufiyaa**
CAPITAL **Malé**	GNP **410 US $ per person**
POPULATION **215,000** Density **721 people per sq km** Life Expectancy **59 years** Infant Mortality **48 per 1000**	MAIN PRODUCTS **Fish,** **Clothing**
	NATIONAL DAY **July 26**
LANGUAGES **Divehi**	ORGANISATIONS **UN, CP, C**

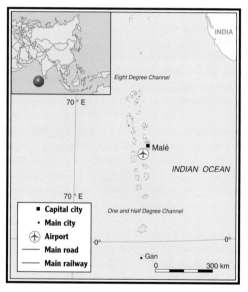

UK PROTECTORATE since 1887; independent 1965. Consists of some 2,000 coral islands stretching across the Indian Ocean; only about 215 are inhabited. Fishing is principal activity, and fish and coconut fibre are both exported. Most staple foods have to be imported but coconuts, millet, cassava, yams and fruit are grown locally. Tourism is developing.

MALI

STATUS **Republic**

AREA **1,240,140 sq km**

CAPITAL **Bamako**

MAIN CITIES **Ségou, Mopti**

POPULATION **8,156,000**
Density **7 people per sq km**
Life Expectancy **46 years**
Infant Mortality **159 per 1000**

LANGUAGES **French, Bambara**

RELIGIONS **Sunni Muslim**

CURRENCY **CFA Franc**

GNP **260 US $ per person**

MAIN PRODUCTS **Cotton, Live animals, Groundnuts**

NATIONAL DAY **September 22**

ORGANISATIONS **UN, OAU, ECOWAS**

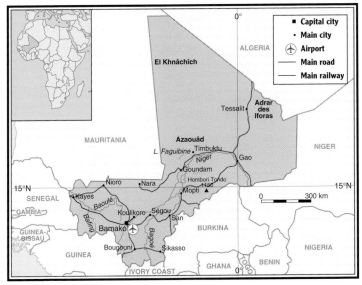

F **RENCH COLONY** of French Sudan since 1904; independent 1960. One of the world's most underdeveloped countries. Most of this landlocked country is barren desert; in the south savannah supports a wide variety of wildlife. Most people live in the Niger valley, growing cotton, oil seeds and groundnuts. Fishing is important. Main exports are cotton and livestock. Recent droughts have taken their toll of livestock and agriculture. There are few mineral resources and no industry.

MALTA

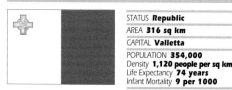

STATUS **Republic**

AREA **316 sq km**

CAPITAL **Valletta**

POPULATION **354,000**
Density **1,120 people per sq km**
Life Expectancy **74 years**
Infant Mortality **9 per 1000**

LANGUAGES **Maltese, English**

RELIGIONS **Roman Catholic**

CURRENCY **Lira**

GNP **5,820 US $ per person**

MAIN PRODUCTS
**Manufactured goods,
Machinery**

NATIONAL DAY **September 21**

ORGANISATIONS **UN, C**

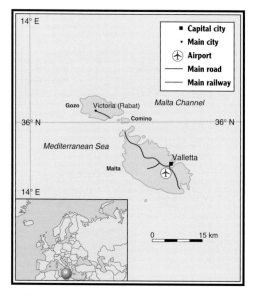

ANNEXED by UK 1814; independent 1964. There are three islands: Malta, Gozo and Comino. Mediterranean climate with mild winters, hot, dry summers and low rainfall. Main crops are wheat, potatoes, tomatoes and vines. Tourism important source of revenue. Principal exports are machinery, beverages, tobacco, flowers, wine, leather goods and potatoes. Large natural harbour at Valletta has made it a major transit port.

MARSHALL ISLANDS

STATUS **Republic**

AREA **181 sq km**

CAPITAL **Dalap-Uliga-Darrit**

POPULATION **40,000**
Density **221 people per sq km**
Life Expectancy **73 years**
Infant Mortality **30 per 1000**

LANGUAGES **Marshallese, English**

RELIGIONS **Protestant**

CURRENCY **US Dollar**

GNP **1,500 US $ per person**

MAIN PRODUCTS **Coconut oil, Fish, Coral, Shells**

ORGANISATIONS **UN, SPF**

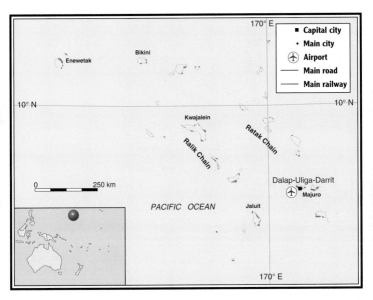

FORMERLY part of the US administered UN Trust Territory of the Pacific Islands. Independent in 1991. Comprises of two long parallel chains of coral atolls. The climate is tropical maritime, with little variation in temperature. Rainfall is heaviest from July to October. The economy is almost totally dependent on US military related payments for use of the islands as bases, although attempts are being made to diversify the economy.

MARTINIQUE see page 91

MAURITANIA

STATUS **Republic**

AREA **1,030,700 sq km**

CAPITAL **Nouakchott**

MAIN CITIES **Nouadhibou**

POPULATION **2,025,000**
Density **2 people per sq km**
Life Expectancy **48 years**
Infant Mortality **117 per 1000**

LANGUAGES **Arabic, French**

RELIGIONS **Sunni Muslim**

CURRENCY **Ouguiya**

GNP **490 US $ per person**

MAIN PRODUCTS **Fish, Iron ore**

NATIONAL DAY **November 28**

ORGANISATIONS **UN, OAU, ECOWAS, AL**

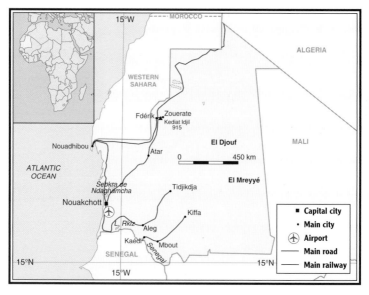

FRENCH PROTECTORATE since 1903; independent 1960. Situated on west coast of Africa. Consists of savannah, arid scrub and desert with high temperatures, low rainfall and frequent droughts. Very little arable farming except in Sénégal River valley, where millet and dates grow. Most people raise cattle, sheep, goats or camels. Severe drought during the last decade decimated livestock population and forced many nomads into the towns. Coastal fishing and fish products earn vital foreign exchange; iron ore and copper are also exported.

STATUS **Constitutional Monarchy**

AREA **1,865 sq km**

CAPITAL **Port Louis**

MAIN CITIES **Beau Bassin-Rose Hill, Curepipe**

POPULATION **1,075,000**
Density **576 people per sq km**
Life Expectancy **70 years**
Infant Mortality **20 per 1000**

LANGUAGES **English**

RELIGIONS **Hindu**

CURRENCY **Rupee**

GNP **1,950 US $ per person**

MAIN PRODUCTS **Clothing, Sugar**

NATIONAL DAY **March 12**

ORGANISATIONS **UN, OAU, PTA, C**

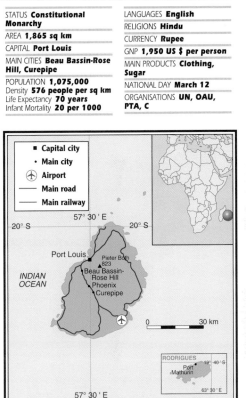

UK POSSESSION since 1810; independent 1968. Mountainous island in Indian Ocean. Temperatures vary from cold to very hot; rainfall from moderate to very high. Sugar-cane and its by-products are mainstay of economy. Tourism is developing rapidly.

MAYOTTE see page 93

MEXICO

STATUS **Federal Republic**

AREA **1,972,545 sq km**

CAPITAL **Mexico City**

MAIN CITIES **Guadalajara, Monterrey, Léon, Ciudad Juárez**

POPULATION **86,154,000**
Density **44 people per sq km**
Life Expectancy **70 years**
Infant Mortality **36 per 1000**

LANGUAGES **Spanish**

RELIGIONS **Roman Catholic**

CURRENCY **Peso**

GNP **1,990 US $ per person**

MAIN PRODUCTS **Crude oil, Metal products, Machinery,**

NATIONAL DAY **September 16**

ORGANISATIONS **UN, OAS, ALADI**

SPANISH colony since 16th century; independent 1821. Landscape consists of mountain ranges and high plateaux. The north is dry but the south is humid and tropical. Population has outstripped food production and many Mexicans have moved to the cities. Minerals, especially silver, uranium and gold, are main source of wealth. Oil, natural gas and coal all have considerable reserves and are becoming more important. Main exports are crude oil and machinery, along with coffee and frozen shrimps. Tourism brings in vital foreign revenue.

MICRONESIA

STATUS **Republic**	RELIGIONS **Christian**
AREA **702 sq km**	CURRENCY **US Dollar**
CAPITAL **Palikir on Pohnpei**	GNP **1,500 US $ per person**
POPULATION **99,000**	MAIN PRODUCTS **Copra,**
Density **141 people per sq km**	**Animal and vegetable oil**
Life Expectancy **73 years**	ORGANISATIONS **UN, SPF**
Infant Mortality **26 per 1000**	
LANGUAGES **English**	

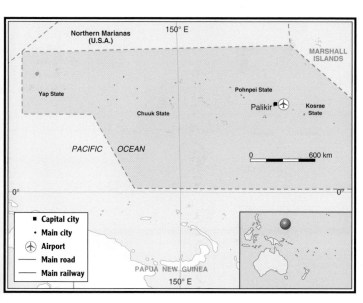

FORMERLY part of the US administered UN Trust Territory of the Pacific Islands, known as the Caroline Islands. Independent 1991. There are over 600 islands in the archipelago, but most are uninhabited. The climate is tropical maritime, with little differences in temperature throughout the year. Rainfall is high all year round, with more falling from July to October. Still closely linked to the US, Micronesia relies heavily on US aid. Attempts are being made to diversify the economy.

MOLDAVIA

STATUS **Republic**

AREA **33,700 sq km**

CAPITAL **Kishinev**

MAIN CITIES **Tiraspol, Beltsy, Bendery**

POPULATION **4,368,000**
Density **130 people per sq km**
Life Expectancy **69 years**
Infant Mortality **21 per 1000**

LANGUAGES **Romanian**

RELIGIONS **Orthodox**

CURRENCY **Leu**

MAIN PRODUCTS **Machinery, Grapes, Consumer products**

ORGANISATIONS **UN**

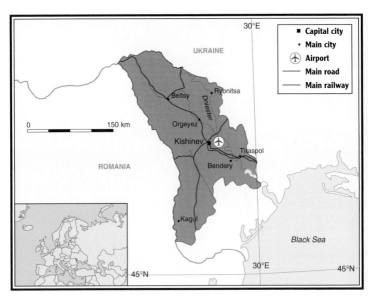

INDEPENDENT 1991. Country of hilly plains. Warm, dry climate with relatively mild winters. Arable farming on very fertile soil; fruit (including grapes for wine), vegetables, sunflower seeds, wheat and maize are chief crops. Industry was dominated by food processing but machine-building and engineering are now expanding. Although Moldavia has close ethnic, linguistic and historic ties with neighbouring Romania, any moves towards re-unification have been fiercely resisted by the Russian minority in the east.

MONACO

STATUS **Constitutional Monarchy**

AREA **2 sq km**

CAPITAL **Monaco-Ville**

POPULATION **29,000**
Density **14,500 people per sq km**
Life Expectancy **76 years**
Infant Mortality **9 per 1000**

LANGUAGES **French**

RELIGIONS **Roman Catholic**

CURRENCY **French Franc**

GNP **11,350 US $ per person**

MAIN PRODUCTS **Chemicals, Plastics**

NATIONAL DAY **November 19**

ORGANISATIONS **UN**

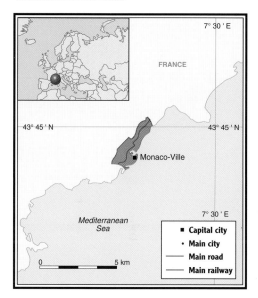

FORMED in 10th century. World's smallest independent state after the Vatican City. Occupies a thin strip of French Mediterranean coast near the Italian border. Most revenue comes from tourism, casinos and light industry. Land has been reclaimed from the sea to extend the area available for commercial development.

MONGOLIA

STATUS **Republic**

AREA **1,565,000 sq km**

CAPITAL **Ulan Bator**

POPULATION **2,190,000**
Density **1 person per sq km**
Life Expectancy **64 years**
Infant Mortality **60 per 1000**

LANGUAGES **Khalka Mongol**

RELIGIONS **Shamanist**

CURRENCY **Tugrik**

GNP **470 US $ per person**

MAIN PRODUCTS **Minerals and fuels, Cattle and horses**

NATIONAL DAY **July 11**

ORGANISATIONS **UN**

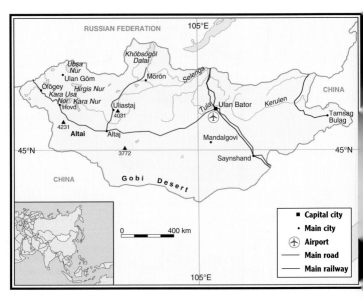

ONE OF THE LOWEST population densities in the world. Much of the country consists of high, undulating plateau covered with grassland. To the north, mountain ranges bridge the border with the Russian Federation; to the south is the vast, arid Gobi Desert. Economy dominated by farming - main exports being cattle and horses. Wheat, barley, millet and oats also grown. Natural resources include oil, coal, iron ore, gold, tin and copper.

MONTSERRAT see page 217

MOROCCO

STATUS **Constitutional Monarchy**

AREA **446,550 sq km**

CAPITAL **Rabat**

MAIN CITIES **Casablanca, Fez, Marrakesh, Meknês**

POPULATION **25,061,000**
Density **56 people per sq km**
Life Expectancy **63 years**
Infant Mortality **68 per 1000**

LANGUAGES **Arabic**

RELIGIONS **Sunni Muslim**

CURRENCY **Dirham**

GNP **900 US $ per person**

MAIN PRODUCTS **Food, Phosphates**

NATIONAL DAY **March 3**

ORGANISATIONS **UN, AL**

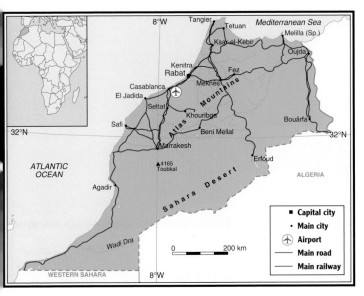

FORMERLY part of Tangier and French and Spanish protectorates; independent 1956. In the north-west are the Atlas Mountains. Between them and the Atlantic Ocean is an area of high plateau bordered on the south by the Sahara Desert. Northern Morocco has a Mediterranean climate and vegetation; rainfall is high on the west-facing, thickly forested slopes of the Atlas Mountains. World's largest phosphate deposits. Main crops are wheat and barley. Tourism is a major industry.

MOZAMBIQUE

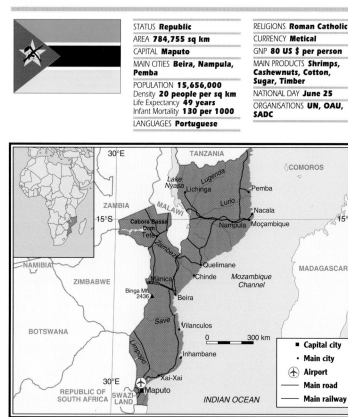

STATUS **Republic**

AREA **784,755 sq km**

CAPITAL **Maputo**

MAIN CITIES **Beira, Nampula, Pemba**

POPULATION **15,656,000**
Density **20 people per sq km**
Life Expectancy **49 years**
Infant Mortality **130 per 1000**

LANGUAGES **Portuguese**

RELIGIONS **Roman Catholic**

CURRENCY **Metical**

GNP **80 US $ per person**

MAIN PRODUCTS **Shrimps, Cashewnuts, Cotton, Sugar, Timber**

NATIONAL DAY **June 25**

ORGANISATIONS **UN, OAU, SADC**

PORTUGUESE colony since 1505; independent 1975. Extensive coastal plain; inland are plateaux with mountain ranges that border Malawi, Zambia and Zimbabwe. Tropical climate on coast; high altitudes make it cooler inland. Most people are subsistence farmers cultivating coconuts, cashews, cotton, maize and rice. Underexploited reserves of coal (considerable), iron ore, bauxite and gold. Acts as trade outlet for Zimbabwe, Zambia and Malawi.

MYANMA see **BURMA** page 56

STATUS **Republic**

AREA **824,295 sq km**

CAPITAL **Windhoek**

POPULATION **1,781,000**
Density **2 people per sq km**
Life Expectancy **59 years**
Infant Mortality **97 per 1000**

LANGUAGES **Afrikaans, English**

RELIGIONS **Protestant**

CURRENCY **SA Rand**

GNP **1,245 US $ per person**

MAIN PRODUCTS **Diamonds, Uranium**

NATIONAL DAY **March 21**

ORGANISATIONS **UN, OAU, SADC, C**

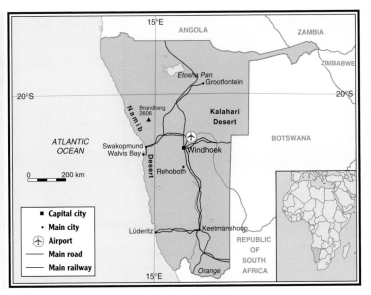

GERMAN protectorate since 1880; administered by South Africa 1915; independent 1990. This south-west African country is one of the driest in the world. Namib Desert on the coast has minimal rainfall; the Kalahari Desert to the north-east has only slightly more rain. Vegetation is sparse. Maize and sorghum grow in the northern highlands; sheep are reared in the south. Rich in mineral resources, with large deposits of diamonds, lead, tin and zinc, and the world's largest uranium mine.

NAURU

STATUS **Republic**	RELIGIONS **Christian**
AREA **21 sq km**	CURRENCY **Australian Dollar**
CAPITAL **Yaren**	GNP **9,091 US $ per person**
POPULATION **10,000**	MAIN PRODUCTS **Phosphates**
Density **476 people per sq km**	NATIONAL DAY **January 31**
Life Expectancy **67 years**	ORGANISATIONS **C, SPF**
Infant Mortality **41 per 1000**	
LANGUAGES **Nauruan, English**	

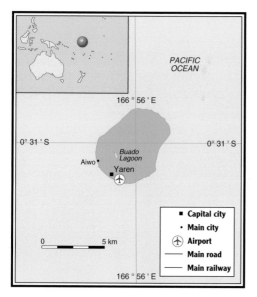

ANNEXED by Germany 1888; Australian UN trusteeship 1941; independent 1968. One of the smallest republics in the world. Great wealth has been derived entirely from phosphate deposits. Flat, coastal lowlands encircled by coral reefs rise gently to the central plateau where the phosphate is mined. Most phosphate is exported to Australia and Japan. Deposits may soon be exhausted.

STATUS **Constitutional Monarchy**

AREA **141,415 sq km**

CAPITAL **Kathmandu**

POPULATION **18,916,000**
Density **134 people per sq km**
Life Expectancy **54 years**
Infant Mortality **118 per 1000**

LANGUAGES **Nepali**

RELIGIONS **Hindu**

CURRENCY **Rupee**

GNP **170 US $ per person**

MAIN PRODUCTS **Carpets, Jute, Handicrafts, Hides and skins**

NATIONAL DAY **February 18**

ORGANISATIONS **UN, CP**

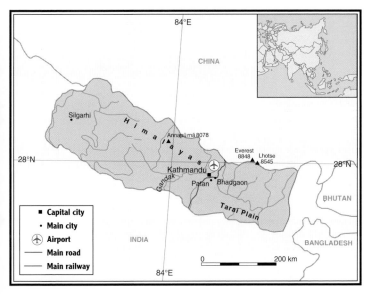

HIMALAYAN kingdom sandwiched between China and India. Climate changes sharply with altitude from the southern Tarai Plain to the northern Himalayas. Most rain falls between June and October. Agriculture concentrated on rice, maize, cattle, buffaloes, sheep and goats. The small amount of industry processes local products.

NETHERLANDS

STATUS **Constitutional Monarchy**

AREA **41,160 sq km**

CAPITAL **Amsterdam**

MAIN CITIES **The Hague (administrative capital), Rotterdam, Utrecht**

POPULATION **14,937,000**
Density **365 people per sq km**
Life Expectancy **78 years**
Infant Mortality **8 per 1000**

LANGUAGES **Dutch**

RELIGIONS **Roman Catholic, Protestant**

CURRENCY **Guilder**

GNP **16,010 US $ per person**

MAIN PRODUCTS **Machinery, Transport equipment, Food stuffs, Beverages**

ORGANISATIONS **UN, EC, OECD, NATO**

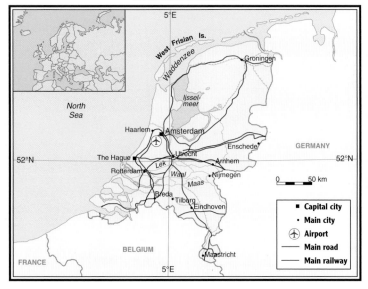

SITUATED at the western edge of the North European Plain. Exceptionally low-lying; complex network of dykes and canals prevents flooding. Mild winters and cool summers. Leading world producers of dairy goods; also crops such as wheat, barley, oats and potatoes. Lacking mineral resources, much industry is dependent on natural gas. Industries such as oil refineries, steel-works and chemical and food processing plants are clustered around Rotterdam.

NETHERLANDS Overseas Territories

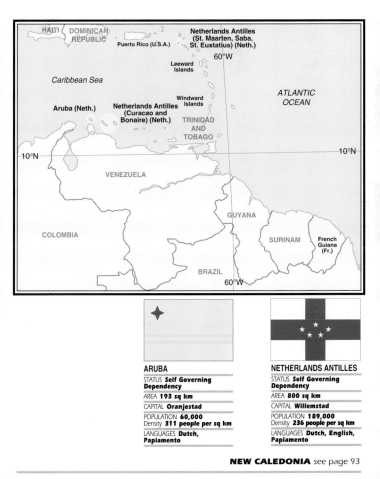

ARUBA

STATUS **Self Governing Dependency**

AREA **193 sq km**

CAPITAL **Oranjestad**

POPULATION **60,000**
Density **311 people per sq km**

LANGUAGES **Dutch, Papiamento**

NETHERLANDS ANTILLES

STATUS **Self Governing Dependency**

AREA **800 sq km**

CAPITAL **Willemstad**

POPULATION **189,000**
Density **236 people per sq km**

LANGUAGES **Dutch, English, Papiamento**

NEW CALEDONIA see page 93

NEW ZEALAND

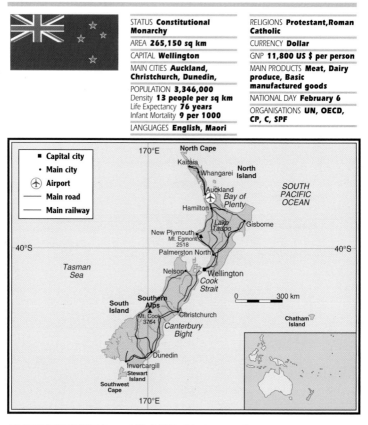

STATUS **Constitutional Monarchy**

AREA **265,150 sq km**

CAPITAL **Wellington**

MAIN CITIES **Auckland, Christchurch, Dunedin,**

POPULATION **3,346,000**
Density **13 people per sq km**
Life Expectancy **76 years**
Infant Mortality **9 per 1000**

LANGUAGES **English, Maori**

RELIGIONS **Protestant, Roman Catholic**

CURRENCY **Dollar**

GNP **11,800 US $ per person**

MAIN PRODUCTS **Meat, Dairy produce, Basic manufactured goods**

NATIONAL DAY **February 6**

ORGANISATIONS **UN, OECD, CP, C, SPF**

- ■ Capital city
- • Main city
- ✈ Airport
- — Main road
- — Main railway

North Cape
Kaitaia
Whangarei · **North Island**
Auckland ✈ · *Bay of Plenty*
Hamilton
New Plymouth · *Lake Taupo* · Gisborne
Mt. Egmont 2518
Palmerston North
Nelson · Wellington ■
Cook Strait
Southern Alps
South Island
Mt. Cook 3764
Christchurch
Canterbury Bight
Dunedin
Invercargill
Stewart Island
Southwest Cape

170°E
40°S

Tasman Sea

SOUTH PACIFIC OCEAN

0 _____ 300 km

Chatham Island

INDEPENDENT from UK 1907. Made up of several islands lying in the South Pacific Ocean. South Island has narrow coastal strip in the west, mountains along centre, and broader plain to the east. North Island is less mountainous. Most of country enjoys temperate climate. Considerable pastureland, some forest. One of world's leading exporters of beef, mutton and wool. Manufacturing and tourism increasing in importance. New trading links are developing with countries bordering the Pacific Ocean.

NEW ZEALAND Overseas Territories

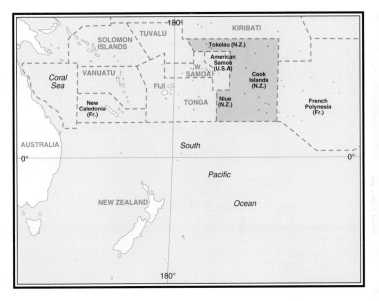

Map of the South Pacific showing:
KIRIBATI, TUVALU, SOLOMON ISLANDS, VANUATU, Coral Sea, New Caledonia (Fr.), AUSTRALIA, FIJI, TONGA, W. SAMOA, American Samoa (U.S.A), Tokelau (N.Z.), Niue (N.Z.), Cook Islands (N.Z.), French Polynesia (Fr.), NEW ZEALAND, South Pacific Ocean

180°, 0°

COOK ISLANDS

STATUS **Self Governing Territory**

AREA **293 sq km**

CAPITAL **Avarua**

POPULATION **18,000**
Density **61 people per sq km**

LANGUAGES **English, Maori**

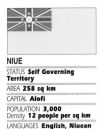

NIUE

STATUS **Self Governing Territory**

AREA **258 sq km**

CAPITAL **Alofi**

POPULATION **3,000**
Density **12 people per sq km**

LANGUAGES **English, Niuean**

TOKELAU

STATUS **Non-Self Governing Territory**

AREA **10 sq km**

POPULATION **2,000**
Density **200 people per sq km**

LANGUAGES **English, Tokelauan**

NICARAGUA

STATUS **Republic**	RELIGIONS **Roman Catholic**
AREA **148,000 sq km**	CURRENCY **Córdoba**
CAPITAL **Managua**	GNP **800 US $ per person**
MAIN CITIES **León, Granada, Chinandega**	MAIN PRODUCTS **Coffee, Cotton, Beef, Bananas, Gold**
POPULATION **3,871,000** Density **26 people per sq km** Life Expectancy **66 years** Infant Mortality **50 per 1000**	NATIONAL DAY **September 15** ORGANISATIONS **UN, OAS, CACM**
LANGUAGES **Spanish**	

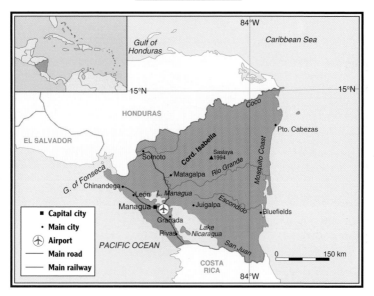

SPANISH colony since 1519; independent 1821. Largest
Central American republic south of Mexico. To the west are
active volcanic mountains which parallel Pacific Ocean coastline.
South is dominated by Lakes Managua and Nicaragua. To the
east is the Caribbean Sea. Tropical climate with rains May to
October. Agriculture is the main occupation; cotton, coffee,
sugar-cane and fruit the leading crops. Gold, silver and copper
are mined.

STATUS **Republic**	RELIGIONS **Sunni Muslim**
AREA **1,186,410 sq km**	CURRENCY **CFA Franc**
CAPITAL **Niamey**	GNP **290 US $ per person**
POPULATION **7,732,000**	MAIN PRODUCTS **Uranium,**
Density **7 people per sq km**	**Live animals**
Life Expectancy **47 years**	NATIONAL DAY **December 18**
Infant Mortality **124 per 1000**	ORGANISATIONS **UN, OAU,**
LANGUAGES **French**	**ECOWAS**

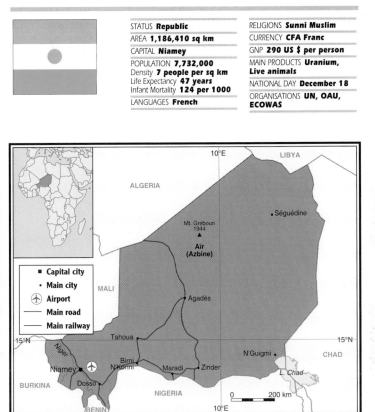

PART OF former French West Africa; independent 1960. Vast, landlocked country south of Sahara Desert. Low rainfall gradually decreases to near zero in the north. Very high temperatures for much of year. Most people farm, particularly cattle, sheep and goats. Recent droughts have affected livestock and cereals. Large deposits of uranium ore and phosphates are being exploited. Economy depends largely on foreign aid.

NIGERIA

STATUS **Federal Republic**	LANGUAGES **English**
AREA **923,850 sq km**	RELIGIONS **Muslim, Christian**
CAPITAL **Abuja**	CURRENCY **Naira**
MAIN CITIES **Lagos, Ibadan, Ogbomosho, Kano, Oshogbo**	GNP **250 US $ per person**
	MAIN PRODUCTS **Crude oil, Palm kernels, Cocoa**
POPULATION **88,500,000** Density **96 people per sq km** Life Expectancy **53 years** Infant Mortality **96 per 1000**	NATIONAL DAY **October 1**
	ORGANISATIONS **UN, OAU, ECOWAS, OPEC, C**

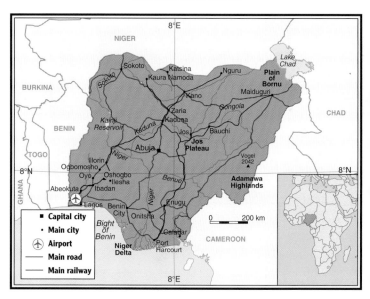

FORMER UK protectorate and colony; independent 1960. Most populous nation in Africa. Bounded to the north by the Sahara Desert, to the west, east and south-east by tropical rainforest. South dominated by the Niger River and its tributaries. Temperatures and humidity are high. From a basic agricultural economy, Nigeria is slowly being transformed by the exploitation of oil in the Niger delta, which accounts for almost all exports.

NIUE see page 157
NORFOLK ISLAND see page 36-37

NORTH KOREA

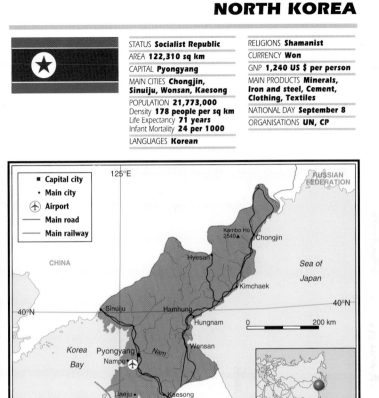

STATUS **Socialist Republic**	RELIGIONS **Shamanist**
AREA **122,310 sq km**	CURRENCY **Won**
CAPITAL **Pyongyang**	GNP **1,240 US $ per person**
MAIN CITIES **Chongjin, Sinuiju, Wonsan, Kaesong**	MAIN PRODUCTS **Minerals, Iron and steel, Cement, Clothing, Textiles**
POPULATION **21,773,000** Density **178 people per sq km** Life Expectancy **71 years** Infant Mortality **24 per 1000**	NATIONAL DAY **September 8**
	ORGANISATIONS **UN, CP**
LANGUAGES **Korean**	

HIGH, RUGGED mountains and deep valleys. Extreme climate with severe winters and warm, sunny summers. Cultivation limited to river valley plains where rice, millet, maize and wheat are principal crops. Rich in minerals including iron ore, coal and copper. Plentiful, underexploited resources of hydro-electricity. Industrial development expanding.

NORTHERN MARIANAS see page 223

NORWAY

STATUS **Constitutional Monarchy**	LANGUAGES **Norwegian**
AREA **323,895 sq km**	RELIGIONS **Protestant**
CAPITAL **Oslo**	CURRENCY **Krone**
MAIN CITIES **Bergen, Trondheim, Stavanger, Kristiansand**	GNP **21,850 US $ per person**
	MAIN PRODUCTS **Crude oil, Natural gas, Metal products, Fish**
POPULATION **4,242,000** Density **13 people per sq km** Life Expectancy **74 years** Infant Mortality **6 per 1000**	NATIONAL DAY **May 17**
	ORGANISATIONS **UN, EFTA, OECD, NATO**

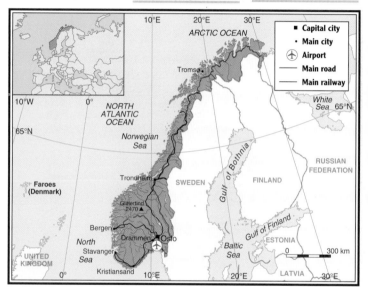

JOINED with Denmark 1397 and Sweden 1814; independent 1905. Mountainous country. Climate modified by Gulf Stream with high rainfall and relatively mild winters. Most people live along fjords, the coast and in the south around Oslo. Rich in natural resources. Advanced production of hydro-electric power has helped develop industry, particularly chemicals, metal products and paper. A leading European oil producer.

STATUS **Monarchy**	RELIGIONS **Muslim**
AREA **271,950 sq km**	CURRENCY **Rial**
CAPITAL **Muscat**	GNP **5,220 US $ per person**
POPULATION **1,502,000**	MAIN PRODUCTS **Crude oil**
Density **6 people per sq km**	NATIONAL DAY **November 18**
Life Expectancy **68 years**	ORGANISATIONS **UN, AL**
Infant Mortality **34 per 1000**	
LANGUAGES **Arabic**	

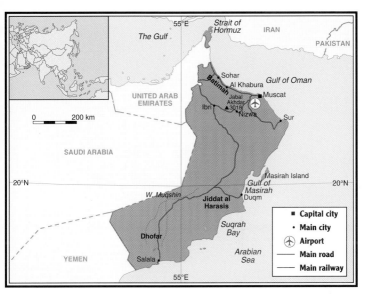

NAME changed from Muscat and Oman 1970. Occupies south-east coast of Arabian Peninsula with a detached portion overlooking the Straits of Hormuz. Desert landscape consists of a coastal plain and low hills rising to interior plateau. Two fertile areas are the Batimah in the north and Dhofar in the south. Main crop is dates. Oil provides almost all export revenue.

PAKISTAN

STATUS **Federal Islamic Republic**	LANGUAGES **Urdu**
AREA **803,940 sq km**	RELIGIONS **Muslim**
CAPITAL **Islamabad**	CURRENCY **Rupee**
MAIN CITIES **Karachi, Lahore, Faisalabad, Rawalpindi**	GNP **370 US $ per person**
POPULATION **112,049,000**	MAIN PRODUCTS **Cotton, Clothing, Leather, Rice, Carpets**
Density **139 people per sq km** Life Expectancy **59 years** Infant Mortality **98 per 1000**	NATIONAL DAY **March 23, August 14**
	ORGANISATIONS **UN, CP, C**

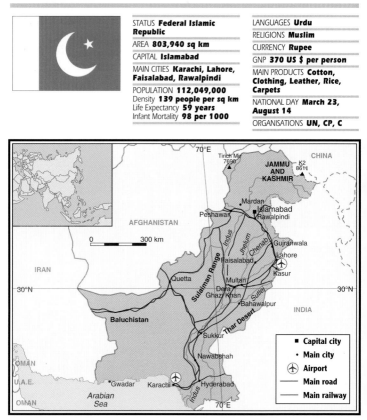

FORMERLY part of British India; independent 1947; separated from Bangladesh (East Pakistan) 1971. Landscape and economy dominated by River Indus and its tributaries, which flow south flanked by plateau of Baluchistan and the Sulaiman Range to the west and the Thar Desert to the east. Dry, hot climate. Farming occurs in irrigated areas near the great rivers. Chief crops are wheat, cotton, maize, rice and sugar-cane. Underdeveloped mineral deposits such as coal and copper. Main industries are food processing and metals.

PALAU see page 223

STATUS **Republic**	RELIGIONS **Roman Catholic**
AREA **78,515 sq km**	CURRENCY **Balboa**
CAPITAL **Panama City**	GNP **1,780 US $ per person**
MAIN CITIES **Colón, David**	MAIN PRODUCTS **Bananas, Shrimps, Clothing, Coffee**
POPULATION **2,418,000**	
Density **31 people per sq km**	NATIONAL DAY **November 3**
Life Expectancy **73 years**	ORGANISATIONS **UN, OAS**
Infant Mortality **21 per 1000**	
LANGUAGES **Spanish**	

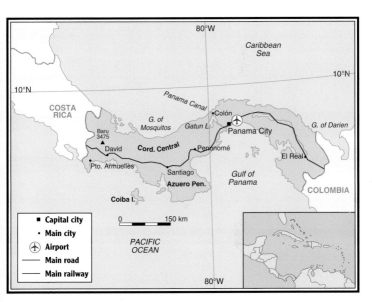

FORMERLY part of Colombia; independent 1903; Canal Zone ceded to US 1903; ceded back to Panama 1979. Situated between the Pacific Ocean and Caribbean Sea at narrowest part of Central America. Tropical climate with little variation in temperature throughout year. Rainy season from April to December. Vast underdeveloped copper reserves, probably largest in world. Most foreign revenue earned from the Panama Canal.

PAPUA NEW GUINEA

STATUS **Constitutional Monarchy**

AREA **462,840 sq km**

CAPITAL **Port Moresby**

POPULATION **3,699,000**
Density **8 people per sq km**
Life Expectancy **56 years**
Infant Mortality **53 per 1000**

LANGUAGES **English**

RELIGIONS **Protestant, Roman Catholic**

CURRENCY **Kina**

GNP **900 US $ per person**

MAIN PRODUCTS **Copper, Gold, Coffee, Timber, Cocoa**

NATIONAL DAY **September 16**

ORGANISATIONS **UN, SPF, CP, C**

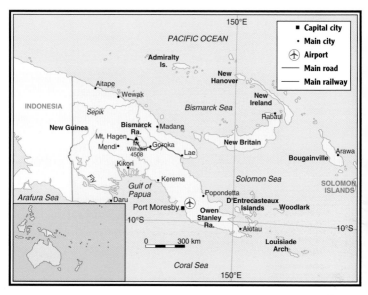

Occupied by Australia 1883; UN trusteeship 1945; independent 1975. Comprises eastern half of New Guinea island and neighbouring islands such as Bougainville. Mountainous terrain with equatorial climate: high temperatures and high rainfall. Copper is main mineral deposit with large reserves on Bougainville. Sugar and beef-cattle are developing areas of production.

STATUS **Republic**	RELIGIONS **Roman Catholic**
AREA **406,750 sq km**	CURRENCY **Guarani**
CAPITAL **Asunción**	GNP **1,030 US $ per person**
POPULATION **4,277,000** Density **11 people per sq km** Life Expectancy **67 years** Infant Mortality **39 per 1000**	MAIN PRODUCTS **Cotton, Soybeans, Beef, Timber, Coffee**
	NATIONAL DAY **May 15**
LANGUAGES **Spanish, Guarani**	ORGANISATIONS **UN, OAS, ALADI**

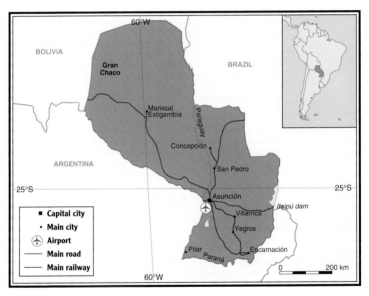

SPANISH possession since 1535; independent 1811. Landlocked country with moderate temperatures all year round. Lush, fertile plains and heavily forested plateaux abound east of the River Paraguay; marshy scrubland (Gran Chaco) lies to the west. Cassava, cotton, soybeans and maize are principal crops. Exports dominated by rearing of livestock - cattle, horses, pigs and sheep - and food processing. World's largest hydro-electric dam, at Itaipú, will eventually have a capacity of 12.6 million kw.

PERU

STATUS **Republic**	RELIGIONS **Roman Catholic**
AREA **1,285,215 sq km**	CURRENCY **Sol**
CAPITAL **Lima**	GNP **1,090 US $ per person**
MAIN CITIES **Arequipa, Trujillo, Callao, Chiclayo**	MAIN PRODUCTS **Copper, Fish, Zinc, Lead**
POPULATION **21,550,000** Density **17 people per sq km** Life Expectancy **65 years** Infant Mortality **76 per 1000**	NATIONAL DAY **July 28**
	ORGANISATIONS **UN, OAS, ALADI**
LANGUAGES **Spanish, Quechua**	

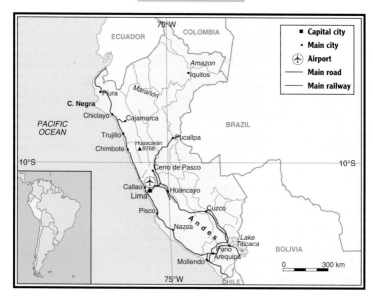

Conquered by Spain early 16th century; independent 1824. Very dry, fertile oases near the coast produce cotton, sugar, fruit and fodder crops. This is the most prosperous, heavily populated area. In the ranges and plateaux of the Andes and lowlands of the Amazon River the soil is thin, yet the inhabitants depend on cultivation and grazing. Rich mineral deposits of copper, lead, zinc and silver. Oil reserves in the interior. Poor communications have hindered development and there are great differences between rich and poor.

STATUS **Republic**	LANGUAGES **Filipino, English**
AREA **300,000 sq km**	RELIGIONS **Roman Catholic**
CAPITAL **Manila**	CURRENCY **Peso**
MAIN CITIES **Quezon City, Davao, Cebu, Bacolod, Zamboanga**	GNP **700 US $ per person**
	MAIN PRODUCTS **Electrical machinery, Clothing, Coconut products**
POPULATION **61,480,000** Density **205 people per sq km** Life Expectancy **65 years** Infant Mortality **40 per 1000**	NATIONAL DAY **June 12**
	ORGANISATIONS **UN, ASEAN, CP**

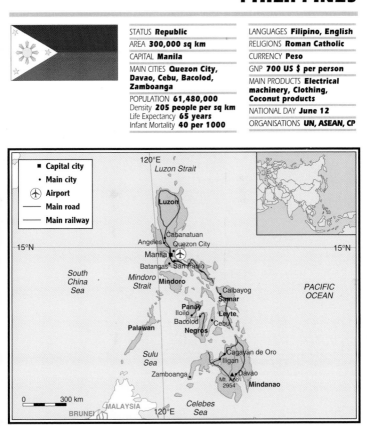

SPANISH colony since 1565; ceded to US 1898; independent 1946. Archipelago comprising three main island groups. Subject to earthquakes and typhoons. Monsoon climate; much of country covered by rainforest. Fishing is important but small farms dominate the economy, producing rice and copra. Main industries are textiles and electrical products. Forestry becoming increasingly important. Unemployment and emigration are both high.

PITCAIRN ISLAND see page 219

POLAND

STATUS **Republic**

AREA **312,685 sq km**

CAPITAL **Warsaw**

MAIN CITIES **Łódź, Kraków, Wrocław, Poznań, Gdańsk**

POPULATION **38,180,000**
Density **122 people per sq km**
Life Expectancy **72 years**
Infant Mortality **17 per 1000**

LANGUAGES **Polish**

RELIGIONS **Roman Catholic**

CURRENCY **Złoty**

GNP **1,760 US $ per person**

MAIN PRODUCTS **Machinery, Transport equipment, Chemicals, Iron and steel**

NATIONAL DAY **May 3**

ORGANISATIONS **UN**

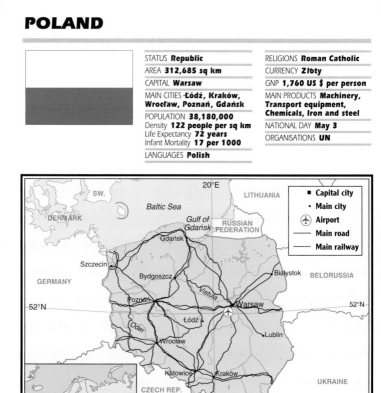

FLAT, WELL-DRAINED North European Plain in north rises gently towards foothills of Carpathian Mountains in the far south. Continental climate with long, severe winters; rainfall is low. Agriculture and natural resources play important part in the economy. Nearly self-sufficient in cereals, sugar-beet and potatoes. Large reserves of coal, copper, sulphur and natural gas. Major industries are ship-building in the north and production of metals and chemicals in major mining centres in the south.

PORTUGAL

STATUS **Parliamentary State**	RELIGIONS **Roman Catholic**
AREA **88,940 sq km**	CURRENCY **Escudo**
CAPITAL **Lisbon**	GNP **4,260 US $ per person**
MAIN CITIES **Porto, Setúbal, Coimbra, Braga**	MAIN PRODUCTS **Textiles, Clothing, Electrical machinery, Footwear, Cork**
POPULATION **10,525,000** Density **118 people per sq km** Life Expectancy **75 years** Infant Mortality **13 per 1000**	NATIONAL DAY **June 10**
LANGUAGES **Portuguese**	ORGANISATIONS **UN, EC, OECD, NATO**

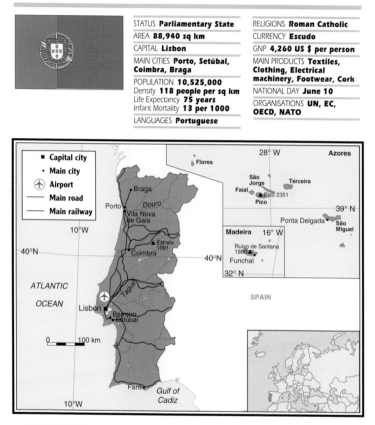

ON THE WEST COAST of Iberian Peninsula, by Atlantic Ocean. Mediterranean climate modified by the Gulf Stream. Lush, mixed deciduous-coniferous forest in the northern mountains; Mediterranean scrub in the far south. One quarter of population are farmers growing vines, olives, wheat, maize and beans. Mineral deposits include coal, copper, kaolinite and uranium. The Azores and Madeira are self governing regions of Portugal; Macau is an overseas territory.

PUERTO RICO see page 222

QATAR

STATUS **Monarchy**	RELIGIONS **Muslim**
AREA **11,435 sq km**	CURRENCY **Riyal**
CAPITAL **Doha**	GNP **9,920 US $ per person**
POPULATION **486,000**	MAIN PRODUCTS **Crude oil, Liquified gas, Chemicals**
Density **43 people per sq km**	
Life Expectancy **70 years**	NATIONAL DAY **September 3**
Infant Mortality **26 per 1000**	ORGANISATIONS **UN, AL, OPEC**
LANGUAGES **Arabic**	

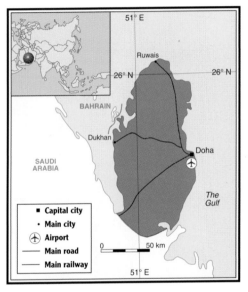

UK **PROTECTORATE** since 1916; independent 1971. Comprises a flat, dry, desert peninsula that reaches north into The Gulf from north-east Saudi Arabia. Climate is hot and humid; rainfall minimal. Irrigation schemes are expanding production of fruit and vegetables for home consumption. Main source of revenue is from exploitation of oil and vast gas reserves.

RÉUNION see page 92

ROMANIA

STATUS **Republic**	LANGUAGES **Romanian**
AREA **237,500 sq km**	RELIGIONS **Orthodox**
CAPITAL **Bucharest**	CURRENCY **Leu**
MAIN CITIES **Braşov, Timişoara, Iaşi, Cluj, Constanţa**	GNP **3,445 US $ per person**
POPULATION **23,200,000**	MAIN PRODUCTS **Machinery, Transport equipment, Fuels, Cement**
Density **96 people per sq km**	
Life Expectancy **72 years**	NATIONAL DAY **August 23**
Infant Mortality **19 per 1000**	ORGANISATIONS **UN**

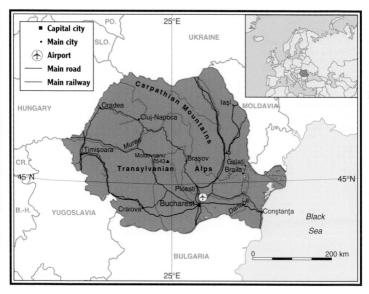

LANDSCAPE dominated by the great curve of the Carpathian Mountains. Lowlands to the west, east and south contain rich agricultural land. Continental climate with variable rainfall, hot summers and cold winters. Forced industrialisation has taken economy from one based on agriculture to one dependent on heavy industry, notably chemicals, metal processing and machine-building. Political and economic prospects for the future look bleak.

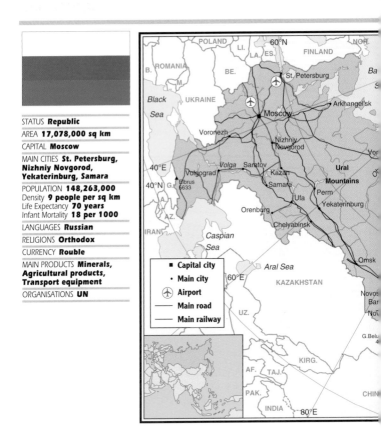

STATUS **Republic**

AREA **17,078,000 sq km**

CAPITAL **Moscow**

MAIN CITIES **St. Petersburg, Nizhniy Novgorod, Yekaterinburg, Samara**

POPULATION **148,263,000**
Density **9 people per sq km**
Life Expectancy **70 years**
Infant Mortality **18 per 1000**

LANGUAGES **Russian**

RELIGIONS **Orthodox**

CURRENCY **Rouble**

MAIN PRODUCTS **Minerals, Agricultural products, Transport equipment**

ORGANISATIONS **UN**

COVERS much of eastern and north-eastern Europe and all of northern Asia. Enormous variation of landforms and climates. Arctic deserts in the north; subtropical semi-deserts in the far south; elsewhere marshy, treeless plains and coniferous forest prevail. Vast natural resources have been key factor in speedy industrialisation during Soviet period. Heavy industry still plays decisive economic role while light and consumer industries are relatively underdeveloped. Future is fraught with political and economic uncertainty.

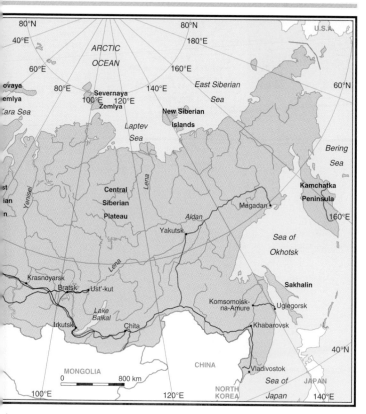

RWANDA

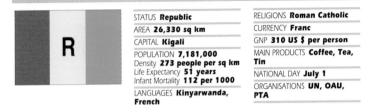

STATUS **Republic**	RELIGIONS **Roman Catholic**
AREA **26,330 sq km**	CURRENCY **Franc**
CAPITAL **Kigali**	GNP **310 US $ per person**
POPULATION **7,181,000** Density **273 people per sq km** Life Expectancy **51 years** Infant Mortality **112 per 1000**	MAIN PRODUCTS **Coffee, Tea, Tin**
	NATIONAL DAY **July 1**
LANGUAGES **Kinyarwanda, French**	ORGANISATIONS **UN, OAU, PTA**

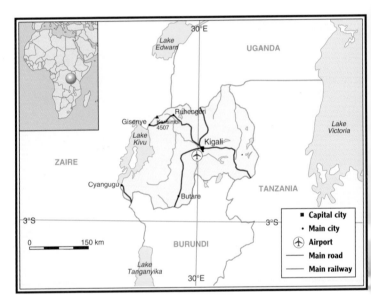

ADMINISTERED by Belgium 1917-62; under UN trusteeship 1946-60; independent 1962. Small, isolated country with moist plateaux east of the Rift Valley supporting dense population. Agriculture is basically subsistence with coffee the major export. Few minerals have been discovered. Manufacturing confined to food processing and construction materials.

ST CHRISTOPHER (ST KITTS)-NEVIS

STATUS **Constitutional Monarchy**

AREA **261 sq km**

CAPITAL **Basseterre**

POPULATION **44,000**
Density **169 people per sq km**
Life Expectancy **68 years**
Infant Mortality **22 per 1000**

LANGUAGES **English**

RELIGIONS **Protestant**

CURRENCY **EC Dollar**

GNP **2,770 US $ per person**

MAIN PRODUCTS **Sugar**

NATIONAL DAY **September 19**

ORGANISATIONS **UN, OAS, CARICOM, C**

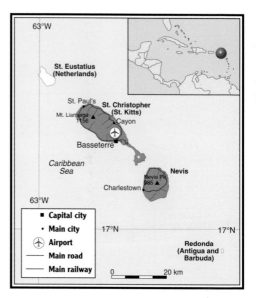

FORMERLY associated state of UK; independent 1983. Comprises two volcanic islands: St Christopher (St Kitts) and Nevis. Climate is tropical and humid with moderate to high temperatures and moderate rainfall. Main crop is sugar. Tourism an important source of revenue.

ST LUCIA

STATUS **Constitutional Monarchy**

AREA **616 sq km**

CAPITAL **Castries**

POPULATION **151,000**
Density **245 people per sq km**
Life Expectancy **71 years**
Infant Mortality **16 per 1000**

LANGUAGES **English**

RELIGIONS **Roman Catholic**

CURRENCY **EC Dollar**

GNP **1,540 US $ per person**

MAIN PRODUCTS **Bananas, Coconut products**

NATIONAL DAY **February 22**

ORGANISATIONS **UN, OAS, CARICOM, C**

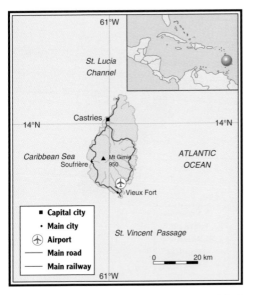

FORMER UK colony; independent 1979. Small, tropical island in the Lesser Antilles in east Caribbean Sea. Most people are small farmers; coconuts, cocoa and bananas the leading crops. Main industries are food and drink processing; tourism is also rapidly developing. There are no commercial mineral deposits and all consumer goods are imported.

ST VINCENT and the GRENADINES

STATUS **Constitutional Monarchy**

AREA **389 sq km**

CAPITAL **Kingstown**

POPULATION **116,000**
Density **298 people per sq km**
Life Expectancy **69 years**
Infant Mortality **23 per 1000**

LANGUAGES **English**

RELIGIONS **Protestant**

CURRENCY **EC Dollar**

GNP **1,100 US $ per person**

MAIN PRODUCTS **Bananas, Arrowroot**

NATIONAL DAY **October 27**

ORGANISATIONS **UN, OAS, CARICOM, C**

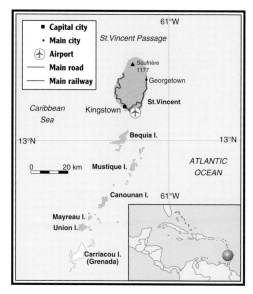

UK COLONY since 1627; changed hands with France several times, restored to UK 1783; independent 1979. Situated in the Lesser Antilles in east Caribbean Sea. Comprises a main island and a chain of small islands called the Northern Grenadines. Climate is tropical. Main crops are arrowroot, sweet potatoes, bananas, coconuts and yams. Some sugar-cane grown for production of rum and other drinks. Tourism is an expanding industry.

ST HELENA AND DEPENDENCIES see page 219
ST PIERRE ET MIQUELON see page 93

SAN MARINO

STATUS **Republic**

AREA **61 sq km**

CAPITAL **San Marino**

POPULATION **24,000**
Density **393 people per sq km**
Life Expectancy **76 years**
Infant Mortality **14 per 1000**

LANGUAGES **Italian**

RELIGIONS **Roman Catholic**

CURRENCY **Italian Lira**

GNP **8,590 US $ per person**

MAIN PRODUCTS **Wine, Wheat, Dairy products, Ceramics**

NATIONAL DAY **September 3**

ORGANISATIONS **UN**

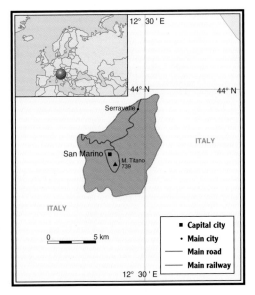

FOUNDED 301. Landlocked country totally surrounded by Italy. Straddles a limestone peak in the Apennines south of Rimini. Economy centred around tourism and sale of postage stamps. Most people are farmers growing cereals, olives and vines and tending sheep and goats.

SAO TOME and PRINCIPE

STATUS **Republic**

AREA **964 sq km**

CAPITAL **São Tomé**

POPULATION **121,000**
Density **126 people per sq km**
Life Expectancy **65 years**
Infant Mortality **70 per 1000**

LANGUAGES **Portuguese**

RELIGIONS **Roman Catholic**

CURRENCY **Dobra**

GNP **280 US $ per person**

MAIN PRODUCTS **Cocoa, Copra**

NATIONAL DAY **July 12**

ORGANISATIONS **UN, OAU**

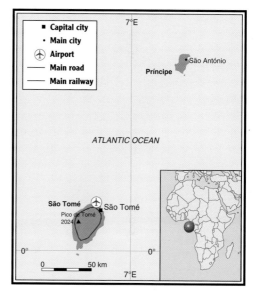

PORTUGUESE overseas province since 1522; independent 1975. Comprises two large islands and several smaller ones. Situated near the equator off the west coast of Africa. Tropical climate with moderately high temperatures and moderate to very high rainfall. Cocoa, coconuts and palm oil are main crops grown on the rich, volcanic soil. Other foods and consumer goods are imported.

181

SAUDI ARABIA

STATUS **Monarchy**	RELIGIONS **Sunni Muslim**
AREA **2,400,900 sq km**	CURRENCY **Riyal**
CAPITAL **Riyadh**	GNP **6,230 US $ per person**
MAIN CITIES **Jeddah, Mecca, Medina, Ta'if, Abha**	MAIN PRODUCTS **Crude oil and petroleum products**
POPULATION **14,870,000** Density **6 people per sq km** Life Expectancy **66 years** Infant Mortality **58 per 1000**	ORGANISATIONS **UN, AL, OPEC**
LANGUAGES **Arabic**	

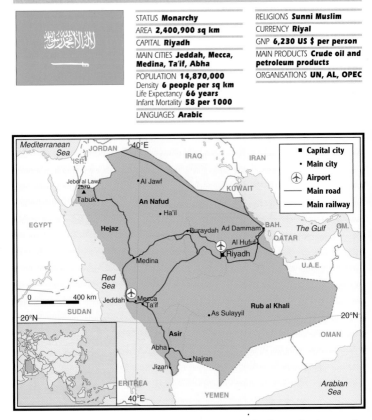

INDEPENDENT after expulsion of Turks 1913. Occupies heart of the vast, arid Arabian Peninsula. Extremely hot in summer. No rivers flow all year round. Interior plateau slopes gently eastwards to The Gulf, supporting little vegetation. Even in the mountainous west, rainfall is low. Only in the coastal strips and oases do cereals and date palms grow. Irrigation schemes and land reclamation projects attempting to raise food production. Economy dominated by oil - Saudi Arabia's most important resource and export commodity.

SENEGAL

STATUS **Republic**	RELIGIONS **Sunni Muslim**
AREA **196,720 sq km**	CURRENCY **CFA Franc**
CAPITAL **Dakar**	GNP **650 US $ per person**
MAIN CITIES **Thiès, Kaolack**	MAIN PRODUCTS **Groundnuts, Fish, Phosphates**
POPULATION **7,327,000**	
Density **37 people per sq km**	NATIONAL DAY **April 4**
Life Expectancy **49 years**	ORGANISATIONS **UN, OAU, ECOWAS**
Infant Mortality **80 per 1000**	
LANGUAGES **French**	

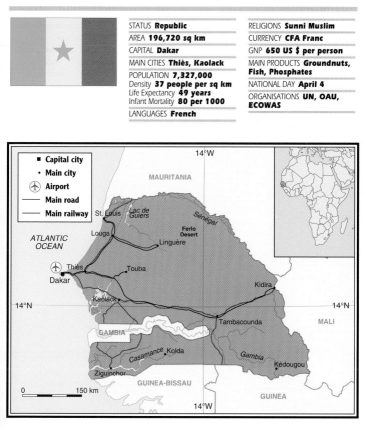

FRENCH colony since 1855; independent 1960; confederation of Senegambia formed with The Gambia 1982. Flat, dry terrain cut through by the Gambia, Casamance and Senegal rivers. Rainfall low, even on coast. Interior, coarse-grassed plain supports varied wildlife but little agriculture. Groundnuts, cotton and millet are main crops, but frequent droughts have reduced their value as cash crops. Phosphate mining, ship-repairing and food processing are the major industries.

SEYCHELLES

STATUS **Republic**
AREA **404 sq km**
CAPITAL **Victoria**
POPULATION **67,000**
Density **166 people per sq km**
Life Expectancy **70 years**
Infant Mortality **13 per 1000**
LANGUAGES **French Creole**

RELIGIONS **Roman Catholic**
CURRENCY **Rupee**
GNP **3,590 US $ per person**
MAIN PRODUCTS **Fish, Copra**
NATIONAL DAY **June 5**
ORGANISATIONS **UN, OAU, C**

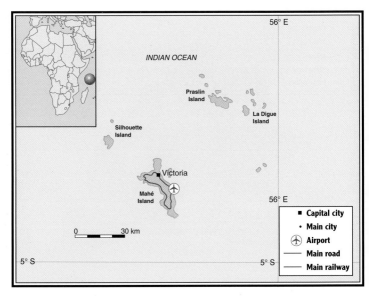

CAPTURED from France by UK 1794; independent 1976. Comprises over 100 granite or coral islands in the Indian Ocean. Temperatures constantly hot; rainfall high to very high. Cassava, sweet potatoes, yams and sugar-cane grown for local consumption; the staple food - rice - is imported. Main exports are copra, coconuts and cinnamon. Fishing also economically important. Tourism has expanded greatly since opening of international airport in 1970s.

SIERRA LEONE

STATUS **Republic**	RELIGIONS **Traditional, Sunni Muslim**
AREA **72,325 sq km**	
CAPITAL **Freetown**	CURRENCY **Leone**
POPULATION **4,151,000**	GNP **200 US $ per person**
Density **57 people per sq km**	MAIN PRODUCTS **Rutile,**
Life Expectancy **43 years**	**Bauxite, Diamonds, Cocoa**
Infant Mortality **143 per 1000**	NATIONAL DAY **April 27**
LANGUAGES **English**	ORGANISATIONS **UN, OAU, ECOWAS, C**

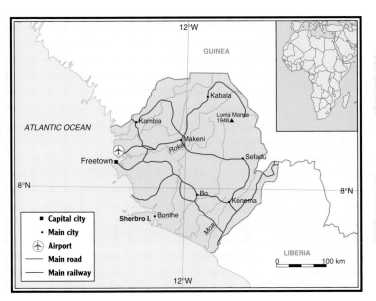

UK COLONY since 1787; independent 1961. Coastline dominated by swamps broken only by small, mountainous peninsula in the west. Wide coastal plain extends inland to foothills of interior plateaux and mountains. Poor, infertile soils; most people subsistence farm. Mineral deposits include diamonds, iron ore and bauxite with manufacturing developed only around the capital.

SINGAPORE

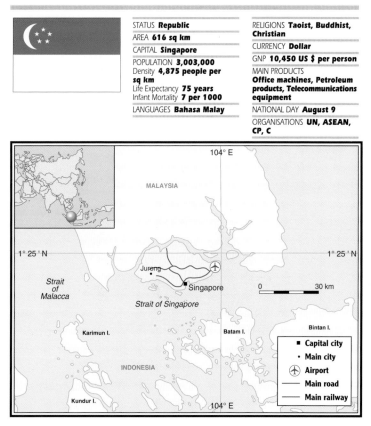

STATUS **Republic**

AREA **616 sq km**

CAPITAL **Singapore**

POPULATION **3,003,000**
Density **4,875 people per sq km**
Life Expectancy **75 years**
Infant Mortality **7 per 1000**

LANGUAGES **Bahasa Malay**

RELIGIONS **Taoist, Buddhist, Christian**

CURRENCY **Dollar**

GNP **10,450 US $ per person**

MAIN PRODUCTS
Office machines, Petroleum products, Telecommunications equipment

NATIONAL DAY **August 9**

ORGANISATIONS **UN, ASEAN, CP, C**

MALAYSIA

104° E

1° 25 ' N

1° 25 ' N

Strait of Malacca

Jureng

Singapore

Strait of Singapore

0   30 km

Karimun I.

Batam I.

Bintan I.

■ **Capital city**

· **Main city**

✈ **Airport**

— **Main road**

— **Main railway**

INDONESIA

Kundur I.

104° E

BRITISH Crown colony 1867; independent 1965. Founded by Sir Stamford Raffles. This mangrove-swamped island has been transformed into one of the world's major entrepreneurial centres. Connected to Peninsular Malaysia by man-made causeway. Hot, humid climate with high rainfall. Few natural resources. Economy dependent on financial services as well as manufacture of precision goods and electronic products.

STATUS **Republic**

AREA **49,036 sq km**

CAPITAL **Bratislavia**

MAIN CITIES **Košice, Banská Bystrica**

POPULATION **5,262,000**
Density **107 people per sq km**
Life Expectancy **71 years**
Infant Mortality **11 per 1000**

LANGUAGES **Slovak**

RELIGIONS **Roman Catholic**

CURRENCY **Koruna**

GNP **3,140 US $ per person**

MAIN PRODUCTS **Machinery, Agricultural products**

ORGANISATIONS **UN**

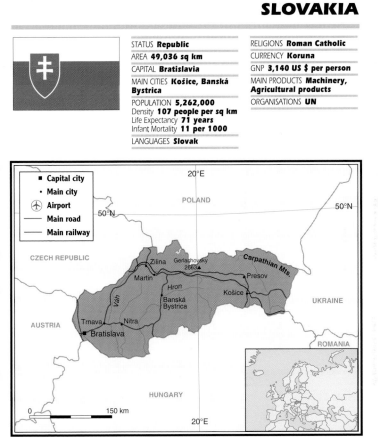

SEPARATED from the Czech Republic 1993. Landlocked country with rugged mountains to the north and agricultural land to the south. Climate is continental with harsh winters and warm summers. Rainfall is low. With the exception of coal, lacks raw materials and minerals. Traditionally has been less developed politically, economically and culturally than the Czech Republic. The communist legacy of inefficient heavy industry has left it ill prepared to face competition, and caused environmental problems.

SLOVENIA

STATUS **Republic**

AREA **20,250 sq km**

CAPITAL **Ljubljana**

MAIN CITIES **Maribor, Celje, Koper**

POPULATION **1,900,000**
Density **94 people per sq km**
Life Expectancy **73 years**
Infant Mortality **9 per 1000**

LANGUAGES **Slovene**

RELIGIONS **Roman Catholic**

CURRENCY **Tolar**

MAIN PRODUCTS **Machinery, Transport equipment, Chemicals**

ORGANISATIONS **UN**

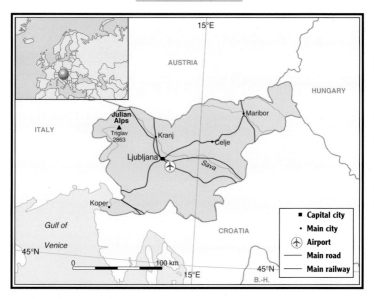

INDEPENDENT from Yugoslavia 1991. One of the key gateways from the Balkans to central and western Europe. Much of country is mountainous with extensive pastures providing profitable dairy-farming; amount of cultivable land restricted. Most people live in the flatter land around the capital. In the north-west are large mercury mines and a broad range of recently developed light industries. Combined with tourism, this has given the country a well-balanced economy.

SOLOMON ISLANDS

STATUS **Constitutional Monarchy**

AREA **29,790 sq km**

CAPITAL **Honiara**

POPULATION **321,000**
Density **11 people per sq km**
Life Expectancy **61 years**
Infant Mortality **74 per 1000**

LANGUAGES **English**

RELIGIONS **Protestant**

CURRENCY **Dollar**

GNP **570 US $ per person**

MAIN PRODUCTS **Timber, Fish, Palm oil, Cocoa**

NATIONAL DAY **July 7**

ORGANISATIONS **UN, C, SPF**

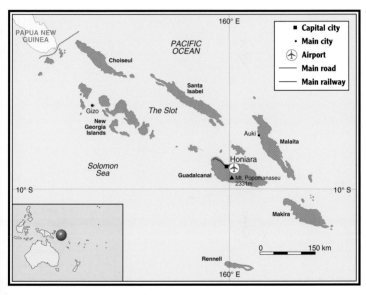

UK **PROTECTORATE** of British Solomon Islands since 1898; independent 1978. Comprises six main and many smaller islands in the South Pacific Ocean. Mountainous, large islands covered by tropical rainforest, reflecting the high temperatures and heavy rainfall. Main crops are coconuts, cocoa and rice; timber, fish and palm oil the principal exports. Reserves of bauxite, phosphate and gold.

SOMALIA

STATUS **Republic**

AREA **630,000 sq km**

CAPITAL **Mogadishu**

MAIN CITIES **Hargeisa, Iscia Baidoa, Kismayu**

POPULATION **7,497,000**
Density **12 people per sq km**
Life Expectancy **47 years**
Infant Mortality **122 per 1000**

LANGUAGES **Arabic, Somali**

RELIGIONS **Sunni Muslim**

CURRENCY **Shilling**

GNP **170 US $ per person**

MAIN PRODUCTS **Live animals, Bananas, Hides and skins**

NATIONAL DAY **October 21**

ORGANISATIONS **UN, OAU, PTA, AL**

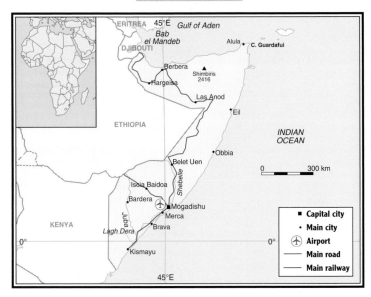

COMBINATION of British Somaliland and Italian Trust Territory of Somalia; independent 1960. Hot, arid country in north-east Africa. Semi-desert of northern mountains contrasts with plains of the south where the bush country is particularly rich in wildlife. Most people are nomads herding camels, sheep, goats and cattle. Little land is cultivated but bananas, cotton, maize, millet and sugar-cane are grown. Underexploited deposits of iron ore, gypsum and uranium. Civil war has disrupted economy.

STATUS **Republic**	LANGUAGES **Afrikaans, English**
AREA **1,184,825 sq km**	RELIGIONS **Christian**
CAPITAL **Cape Town and Pretoria**	CURRENCY **Rand**
	GNP **2,460 US $ per person**
MAIN CITIES **Johannesburg, Durban, Port Elizabeth, Bloemfontein**	MAIN PRODUCTS **Gold, Metal products, Precious stones, Food**
POPULATION **35,282,000** Density **30 people per sq km** Life Expectancy **63 years** Infant Mortality **62 per 1000**	NATIONAL DAY **May 31**
	ORGANISATIONS **UN**

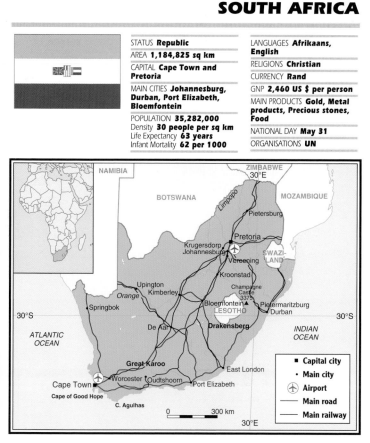

UNION of South Africa 1910; became a republic and withdrew from Commonwealth 1961. Most highly developed African country. Interior plateau, drained by the Orange and Limpopo rivers, is surrounded by escarpment below which land descends to the sea. Sheep and cattle widely grazed. Major crops are maize, wheat, sugar-cane, vegetables, cotton and vines. Abundant reserves of diamonds, platinum, silver, gold, uranium, copper, manganese, asbestos and coal.

SOUTH KOREA

STATUS **Republic**	RELIGIONS **Buddhist, Christian**
AREA **98,445 sq km**	CURRENCY **Won**
CAPITAL **Seoul**	GNP **4,400 US $ per person**
MAIN CITIES **Pusan, Taegu, Inchon, Kwangchu, Taejon**	MAIN PRODUCTS **Machinery, Transport equipment, Chemicals**
POPULATION **42,793,000** Density **435 people per sq km** Life Expectancy **71 years** Infant Mortality **21 per 1000**	NATIONAL DAY **August 15**
	ORGANISATIONS **UN, CP**
LANGUAGES **Korean**	

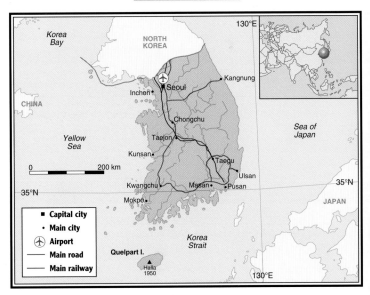

MOUNTAINOUS in east and centre. Most people live in arable river valleys and along coastal plain. Hot, humid summers; dry winters. Agriculture very primitive, with rice the principal crop. Main mineral deposits are tungsten, coal and iron ore. Major industrial nation with iron and steel, chemicals, machinery, ship-building, vehicles and electronics dominating. Builds more ships than any other nation except Japan. Oil and industrial materials have to be imported.

SPAIN

STATUS **Constitutional Monarchy**	LANGUAGES **Spanish**
AREA **504,880 sq km**	RELIGIONS **Roman Catholic**
CAPITAL **Madrid**	CURRENCY **Peseta**
	GNP **9,150 US $ per person**
MAIN CITIES **Barcelona, Valencia, Seville, Zaragoza, Malaga**	MAIN PRODUCTS **Motor vehicles, Petroleum products, Iron ore, Food**
POPULATION **38,959,000** Density **77 people per sq km** Life Expectancy **77 years** Infant Mortality **9 per 1000**	NATIONAL DAY **October 12**
	ORGANISATIONS **UN, EC, OECD, NATO**

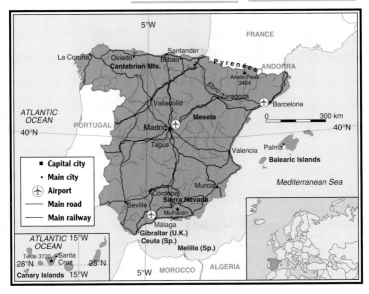

MOUNTAIN ranges fringe vast central plateau. Climate varies according to latitude and closeness to Atlantic Ocean and Mediterranean Sea. Much of land covered by Mediterranean scrub but wheat, barley, maize, grapes and olives are grown. Principal cash crops are cotton, olives, tobacco and citrus fruit. Major industries are textiles, steel, chemicals, consumer goods, vehicle manufacturing, cement, fishing and forestry. Main minerals are coal, iron ore, uranium and zinc. Tourism is important. The Canary Islands are a province of Spain.

SRI LANKA

STATUS **Republic**

AREA **65,610 sq km**

CAPITAL **Colombo**

MAIN CITIES **Dehiwela-Mt.Lavinia, Moratuwa, Jaffna, Kandy**

POPULATION **16,993,000**
Density **259 people per sq km**
Life Expectancy **72 years**
Infant Mortality **24 per 1000**

LANGUAGES **Sinhalese, Tamil**

RELIGIONS **Buddhist, Hindu**

CURRENCY **Rupee**

GNP **430 US $ per person**

MAIN PRODUCTS **Tea, Rubber, Precious stones, Coconut, Cinnamon**

NATIONAL DAY **February 4**

ORGANISATIONS **UN, CP, C**

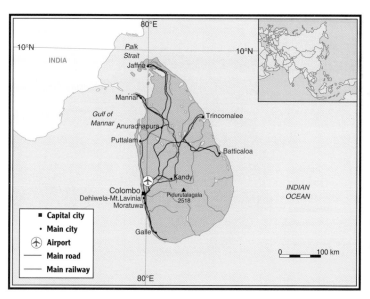

PORTUGUESE colony 1505-1658; Dutch colony 1658-1796; ceded to UK 1798; independent 1948; name changed from Ceylon 1972. Undulating coastal plain encircles central highlands. Climate is tropical on the coast, temperate in the hills. Moderate rainfall in north and east; high in south and west. Rich agricultural land on which tea, rubber and coconuts are cultivated. Gemstones (sapphire, ruby, beryl, topaz), graphite and salt mined. Main industries are food processing, textiles, chemicals and rubber.

STATUS **Military Regime**	LANGUAGES **Arabic**
AREA **2,505,815 sq km**	RELIGIONS **Sunni Muslim**
CAPITAL **Khartoum**	CURRENCY **Pound**
MAIN CITIES **Khartoum North, Omdurman, Port Sudan, Wad Medani**	GNP **540 US $ per person**
	MAIN PRODUCTS **Cotton, Sesame seeds, Gum arabic, Sorghum**
POPULATION **25,204,000** Density **10 people per sq km** Life Expectancy **52 years** Infant Mortality **99 per 1000**	NATIONAL DAY **January 1**
	ORGANISATIONS **UN, OAU, AL**

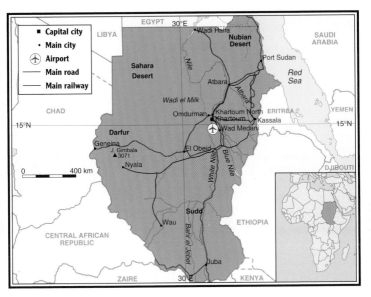

CONDOMINIUM of UK and Egypt since 1899; independent 1956. Africa's largest country. Lies in upper Nile basin. Mostly flat, infertile terrain with hot, arid climate. White and Blue Niles irrigate cultivated land and are potential sources of hydro-electric power. Subsistence farming accounts for much of Sudan's agriculture. Principal activity is nomadic herding of cattle, sheep and goats.

SURINAM

STATUS **Republic**

AREA **163,820 sq km**

CAPITAL **Paramaribo**

POPULATION **422,000**
Density **3 people per sq km**
Life Expectancy **70 years**
Infant Mortality **28 per 1000**

LANGUAGES **Dutch, English**

RELIGIONS **Hindu, Roman Catholic**

CURRENCY **Guilder**

GNP **3,020 US $ per person**

MAIN PRODUCTS **Bauxite, Rice, Shrimps, Bananas**

NATIONAL DAY **November 25**

ORGANISATIONS **UN, OAS**

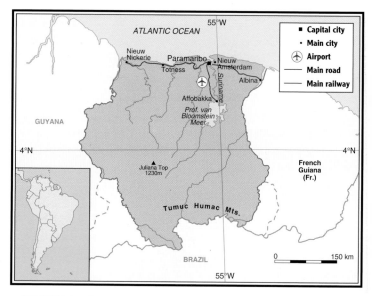

CEDED by Britain to The Netherlands 1667 in return for New York (New Amsterdam); name changed from Dutch Guiana; independent 1975. Lies on north-east coast in the tropics of South America. Low-lying, marshy coastal strip gives way to undulating, coarse grasslands, then densely forested highlands in the south. Timber products and rice are important, the latter being grown on most cultivated land. Introduction of cattle-raising for meat and dairy products is not yet complete. Chief export is bauxite. Timber resources largely untapped.

SWAZILAND

STATUS **Monarchy**

AREA **17,365 sq km**

CAPITAL **Mbabane**

POPULATION **768,000**
Density **44 people per sq km**
Life Expectancy **58 years**
Infant Mortality **107 per 1000**

LANGUAGES **English, Siswati**

RELIGIONS **Christian**

CURRENCY **Lilangeni**

GNP **900 US $ per person**

MAIN PRODUCTS **Sugar, Wood products, Canned fruit and juices**

NATIONAL DAY **September 6**

ORGANISATIONS **UN, OAU, PTA, SADC, C**

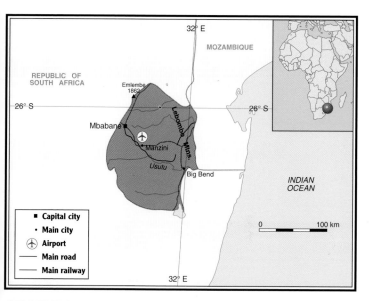

FORMER UK protectorate; independent 1968. Subtropical grassland occurs in three regions: at high altitude in the west, medium in the centre, and low in the east. Lebombo Mountains rise in the far east. Abundant rainfall promotes good pastureland for cattle and sheep.

SWEDEN

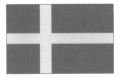

STATUS **Constitutional Monarchy**

AREA **449,790 sq km**

CAPITAL **Stockholm**

MAIN CITIES **Göteborg, Malmö, Uppsala, Linköping, Örebro**

POPULATION **8,559,000**
Density **19 people per sq km**
Life Expectancy **78 years**
Infant Mortality **6 per 1000**

LANGUAGES **Swedish**

RELIGIONS **Protestant**

CURRENCY **Krona**

GNP **21,710 US $ per person**

MAIN PRODUCTS **Transport equipment, Paper products, Electrical machinery**

NATIONAL DAY **June 6**

ORGANISATIONS **UN, EFTA, OECD**

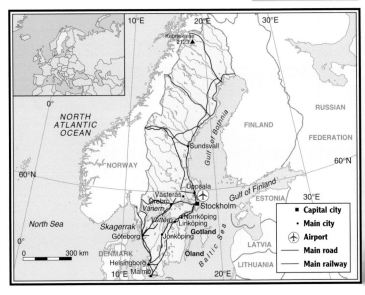

FORESTED mountains in the north give way to central lowlands and lakes, then southern uplands of Jönköping and extremely fertile Scania Plain in the far south. Glacial debris, glacier-eroded valleys and thick glacial clay are also dominant features. Summers are short and hot; winters long and cold. Rainfall varies from moderately heavy to low. Timber industry is important as is manufacturing industry, particularly cars and trucks, metal products and machine tools. Plentiful mineral resources of iron ore, copper, lead and zinc.

SWITZERLAND

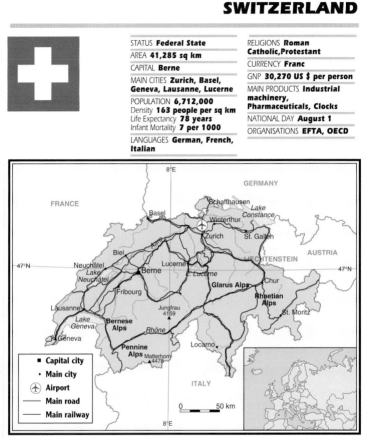

STATUS **Federal State**

AREA **41,285 sq km**

CAPITAL **Berne**

MAIN CITIES **Zurich, Basel, Geneva, Lausanne, Lucerne**

POPULATION **6,712,000**
Density **163 people per sq km**
Life Expectancy **78 years**
Infant Mortality **7 per 1000**

LANGUAGES **German, French, Italian**

RELIGIONS **Roman Catholic, Protestant**

CURRENCY **Franc**

GNP **30,270 US $ per person**

MAIN PRODUCTS **Industrial machinery, Pharmaceuticals, Clocks**

NATIONAL DAY **August 1**

ORGANISATIONS **EFTA, OECD**

MOUNTAINOUS, landlocked country in the Alps. Very cold winters; mild, wet summers. Agriculture based mainly on dairy farming. Major crops include hay, wheat, barley and potatoes. Economically important industry centred on metal engineering, watchmaking, food processing, textiles and chemicals. Tourism and financial services, especially banking, are also important sources of income and employment. Attractive location for international organisations owing to its neutrality during armed conflicts.

SYRIA

STATUS **Republic**	RELIGIONS **Sunni Muslim**
AREA **185,680 sq km**	CURRENCY **Pound**
CAPITAL **Damascus**	GNP **1,100 US $ per person**
MAIN CITIES **Aleppo, Homs, Latakia**	MAIN PRODUCTS **Crude oil, Petroleum products, Textiles, Leather**
POPULATION **12,116,000** Density **65 people per sq km** Life Expectancy **67 years** Infant Mortality **39 per 1000**	NATIONAL DAY **April 17**
	ORGANISATIONS **UN, AL**
LANGUAGES **Arabic**	

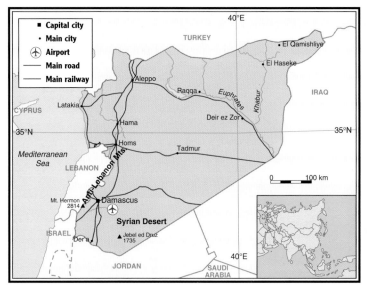

FRENCH mandate since 1920; republic 1941; independent 1944. Most fertile areas lie along Mediterranean Sea, and in depressions and plateaux of the north-east. Anti-Lebanon Mountains in south-west; hot, arid Syrian Desert in the south. Cotton, wheat and barley are grown. Chief livestock are cattle, sheep and goats. Economy was dominated by agriculture but rapid industrialisation now occurring as oil, natural gas and phosphate resources are exploited. Mining of salt and gypsum.

STATUS **Republic**

AREA **35,990 sq km**

CAPITAL **Taipei**

MAIN CITIES **Kaohsiung, Taichung, Keelung**

POPULATION **20,300,000**
Density **564 people per sq km**
Life Expectancy **74 years**
Infant Mortality **6 per 1000**

LANGUAGES **Mandarin**

RELIGIONS **Buddhist, Taoist**

CURRENCY **Dollar**

GNP **7,990 US $ per person**

MAIN PRODUCTS **Electronic equipment, Clothing, Footwear, Plastics**

NATIONAL DAY **October 10**

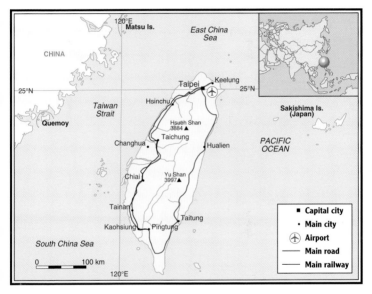

POLITICALLY separated from China 1949; expelled from UN 1971. Extensive mountainous terrain. Most people live and work in the flat to rolling coastal plain in the west. Tropical, marine climate with persistent cloud. Monsoon rainfall from June to August. Main crops are rice, tea, fruit, sugar-cane and sweet potatoes. Light and heavy industry includes textiles and electrical goods. Natural resources of limestone, marble, asbestos, copper and sulphur. Natural gas is extracted from Taiwan Strait.

TAJIKISTAN

STATUS **Republic**

AREA **143,100 sq km**

CAPITAL **Dushanbe**

MAIN CITIES **Khudzand, Kurgan-Tyube, Kulyab**

POPULATION **5,303,000**
Density **37 people per sq km**
Life Expectancy **70 years**
Infant Mortality **43 per 1000**

LANGUAGES **Tajik**

RELIGIONS **Sunni Muslim**

CURRENCY **Rouble**

MAIN PRODUCTS **Horticulture, Cattle, Minerals**

ORGANISATIONS **UN**

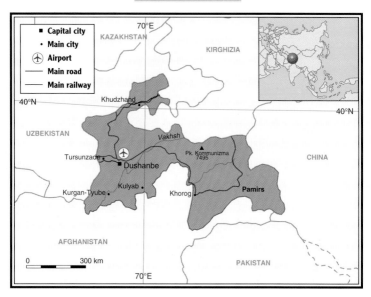

INDEPENDENT 1991. Extremely mountainous terrain. Most people live along major rivers or where mountains join semi-arid plains. Climate varies from continental to subtropical according to height and shelter. Agriculture would be severely limited without extensive irrigation. Leading crop is cotton. Textiles now largest industry. Rich in minerals and fuel deposits, exploitation of which was a major feature of economic development during Soviet era. Recent upsurge of violent nationalism has encouraged foreigners to emigrate.

STATUS **Republic**	RELIGIONS **Christian, Muslim**
AREA **939,760 sq km**	CURRENCY **Shilling**
CAPITAL **Dodoma**	GNP **120 US $ per person**
MAIN CITIES **Dar es Salaam, Zanzibar, Mwanza**	MAIN PRODUCTS **Coffee, Cotton, Sisal**
POPULATION **25,635,000** Density **27 people per sq km** Life Expectancy **55 years** Infant Mortality **97 per 1000**	NATIONAL DAY **April 26**
	ORGANISATIONS **UN, OAU, PTA, SADC, C**
LANGUAGES **Swahili, English**	

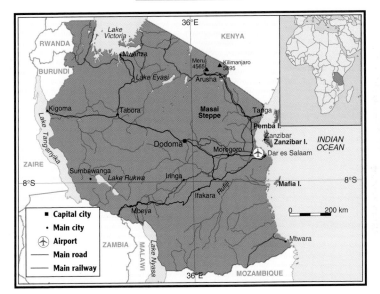

U K TRUST territory of Tanganyika since 1918; independent 1961; union with Zanzibar to form Tanzania 1964. Consists of high interior plateau covered by scrub and grassland, bordered to the north by volcanic Kilimanjaro region, to the east by Lake Tanganyika, and by highlands to the south. Truly tropical climate only on coast. Most people subsistence farm; coffee, cotton, sisal, cashew nuts and tea are exported. Industry, mainly textiles, food processing and tobacco, gradually growing in importance. Tourism could be future growth area.

THAILAND

STATUS **Constitutional Monarchy**

AREA **514,000 sq km**

CAPITAL **Bangkok**

MAIN CITIES **Chiang Mai, Thonburi, Nakhon Ratchasima**

POPULATION **57,196,000**
Density **111 people per sq km**
Life Expectancy **67 years**
Infant Mortality **24 per 1000**

LANGUAGES **Thai**

RELIGIONS **Buddhist**

CURRENCY **Baht**

GNP **1,170 US $ per person**

MAIN PRODUCTS **Electrical machinery, Textiles, Clothing, Fish**

NATIONAL DAY **December 5**

ORGANISATIONS **UN, ASEAN, CP**

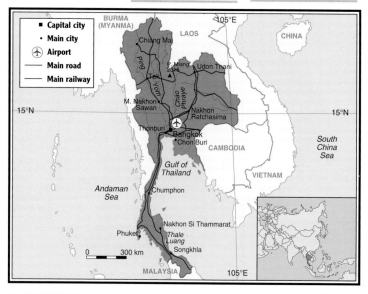

FORMERLY Siam. Flat, undulating central plain containing Chao Phraye River and fringed by mountains; plateau in the northeast drained by Mekong River. From May to October monsoon rains are heavy. Tropical climate with high temperatures. Much of country covered by dense rainforest. Central plain well-served with irrigation canals, which supply paddy fields. Rice, maize, beans, coconuts and groundnuts are main crops. One of world's largest producers of rubber and tin. Small-scale petrochemical industry has been developed.

TOGO

STATUS **Republic**	RELIGIONS **Traditional**
AREA **56,785 sq km**	CURRENCY **CFA Franc**
CAPITAL **Lomé**	GNP **390 US $ per person**
POPULATION **3,531,000** Density **62 people per sq km** Life Expectancy **55 years** Infant Mortality **85 per 1000**	MAIN PRODUCTS **Phosphates, Coffee, Cotton, Cocoa**
	NATIONAL DAY **January 13**
LANGUAGES **French**	ORGANISATIONS **UN, OAU, ECOWAS**

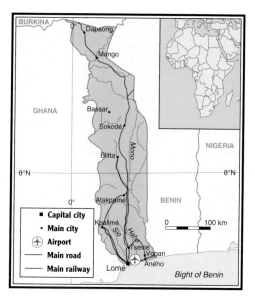

PART OF former German colony of Togoland; administered by France since 1914; independent 1960. Situated between Ghana and Benin on west African coast. Interior comprises mountains and high infertile tableland. Tropical, moderately hot climate. Most farmers grow maize, cassava, yams, groundnuts and plantains. Virtually self-sufficient in food-stuffs. Phosphates account for half of export revenue. Cotton, cocoa and coffee also exported.

TONGA

STATUS **Constitutional Monarchy**

AREA **699 sq km**

CAPITAL **Nuku'alofa**

POPULATION **95,000**
Density **136 people per sq km**
Life Expectancy **66 years**
Infant Mortality **49 per 1000**

LANGUAGES **English, Tongan**

RELIGIONS **Protestant**

CURRENCY **Pa'anga**

GNP **800 US $ per person**

MAIN PRODUCTS **Squash, Vanilla, Root crops, Fish, Coconuts**

NATIONAL DAY **June 4**

ORGANISATIONS **SPF, C**

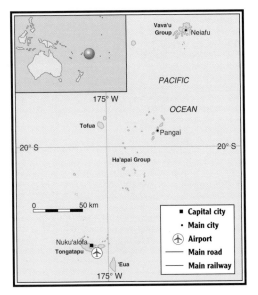

BRITISH protectorate since 1900; independent 1970. Archipelago of 169 islands in Pacific Ocean north of New Zealand. Comprises seven groups of islands but most important are Tongatapu, Ha'apai and Vava'u. All islands covered with dense tropical vegetation. Moderate to moderately high temperatures.

TOKELAU see page 157

TRINIDAD and TOBAGO

STATUS **Republic**

AREA **5,130 sq km**

CAPITAL **Port of Spain**

POPULATION **1,227,000**
Density **239 people per sq km**
Life Expectancy **72 years**
Infant Mortality **14 per 1000**

LANGUAGES **English**

RELIGIONS **Roman Catholic, Hindu**

CURRENCY **Dollar**

GNP **3,160 US $ per person**

MAIN PRODUCTS **Crude oil, Petroleum products, Chemical products**

NATIONAL DAY **August 31, September 24**

ORGANISATIONS **UN, OAS, CARICOM, C**

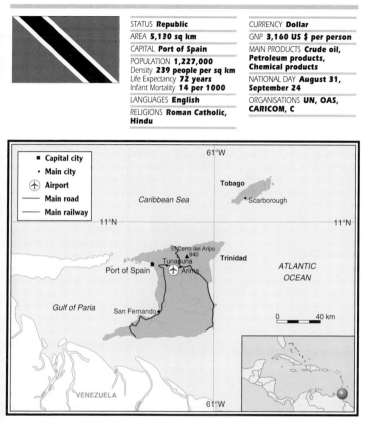

OCCUPIED by UK 1797; islands united 1888; independent 1962. Situated off Venezuelan coast in Caribbean Sea. Both islands have mountainous interiors and are densely covered with tropical rainforest. Sugar was once mainstay of economy but oil is now the leading source of revenue.

TUNISIA

STATUS **Republic**

AREA **164,150 sq km**

CAPITAL **Tunis**

MAIN CITIES **Sfax, Bizerta, Sousse, Gabès**

POPULATION **8,180,000**
Density **50 people per sq km**
Life Expectancy **68 years**
Infant Mortality **44 per 1000**

LANGUAGES **Arabic**

RELIGIONS **Sunni Muslim**

CURRENCY **Dinar**

GNP **1,260 US $ per person**

MAIN PRODUCTS **Clothing, Crude oil, Phosphates, Fish**

NATIONAL DAY **March 20**

ORGANISATIONS **UN, OAU, AL**

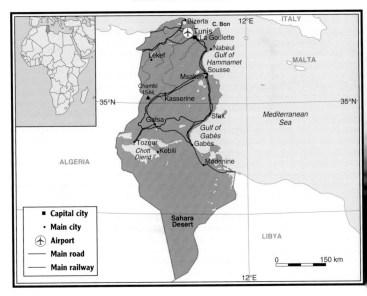

FRENCH protectorate since 1881; independent 1956. Fringed to the north and west by eastern end of Atlas Mountains. Salt lakes scattered throughout arid central plains; to the south lies the Sahara Desert. Mediterranean scrub on coastal areas. Most people live along the north-east coast. Wheat, barley, olives and citrus fruit are main crops; oil, natural gas and sugar refining main industries. Tourist industry expanding and becoming increasingly important to economy.

TURKEY

STATUS **Republic**	RELIGIONS **Sunni Muslim**
AREA **779,450 sq km**	CURRENCY **Lira**
CAPITAL **Ankara**	GNP **1,360 US $ per person**
MAIN CITIES **Istanbul, Izmir, Adana, Bursa, Gaziantep**	MAIN PRODUCTS **Textiles, Iron ore, Fruit, Leather**
POPULATION **56,098,000** Density **72 people per sq km** Life Expectancy **66 years** Infant Mortality **62 per 1000**	NATIONAL DAY **October 29**
	ORGANISATIONS **UN, OECD, NATO**
LANGUAGES **Turkish**	

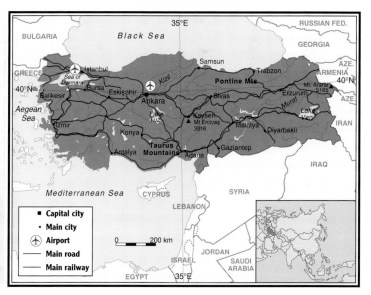

OCCUPIES strategically important position linking Europe and Asia. Central plateau bordered north and south by mountains. North, south and west coastlines fringed by Mediterranean vegetation; short, mild, wet winters and long, hot summers. Low rainfall in arid interior. Main crops are wheat, barley, tobacco, olives, sugar beet, tea and fruit. Sheep, goats and cattle raised. Leading Middle Eastern producer of iron, steel, chrome, coal and lignite. Tourism growing.

TURKMENISTAN

STATUS **Republic**

AREA **488,100 sq km**

CAPITAL **Ashkhabad**

MAIN CITIES **Chardzhou, Mary, Nebit-Dag, Krasnovodsk**

POPULATION **3,670,000**
Density **8 people per sq km**
Life Expectancy **65 years**
Infant Mortality **55 per 1000**

LANGUAGES **Turkmenian**

RELIGIONS **Sunni Muslim**

CURRENCY **Rouble**

MAIN PRODUCTS **Cotton, Wool, Carpets, Minerals**

ORGANISATIONS **UN**

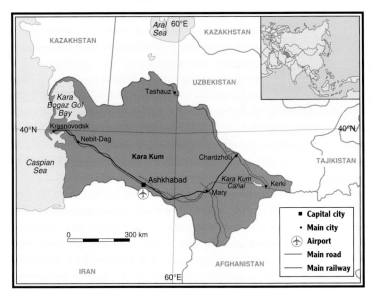

INDEPENDENT 1991. Situated in far south of former USSR. Mainly deserts and oases; only in south do hills and mountains emerge. Continental climate with great temperature fluctuations. Introduction of extensive irrigation during Soviet era has encouraged nomadic pastoralists to switch to cotton growing. Rich in minerals and chemicals. Industrial growth forcibly developed by Russians, resulting in ethnic-outsiders now making up majority of urban population.

TURKS AND CAICOS ISLANDS see page 217

TUVALU

STATUS **Constitutional Monarchy**

AREA **25 sq km**

CAPITAL **Fongafale**

POPULATION **10,000**
Density **400 people per sq km**
Life Expectancy **62 years**
Infant Mortality **33 per 1000**

LANGUAGES **English, Tuvaluan**

RELIGIONS **Protestant**

CURRENCY **Dollar**

GNP **500 US $ per person**

MAIN PRODUCTS **Copra**

NATIONAL DAY **October 1**

ORGANISATIONS **SPF, C**

PART of the former Gilbert and Ellice Islands Colony; separation 1975; independent 1978. Consists of nine dispersed coral atolls, north of Fiji, in Pacific Ocean. Hot, tropical climate with heavy annual rainfall. Fish is staple food. Coconuts and breadfruit cultivated.

UGANDA

STATUS **Republic**	RELIGIONS **Roman Catholic**
AREA **236,580 sq km**	CURRENCY **Shilling**
CAPITAL **Kampala**	GNP **250 US $ per person**
POPULATION **18,795,000** Density **79 people per sq km** Life Expectancy **53 years** Infant Mortality **94 per 1000**	MAIN PRODUCTS **Coffee, Cotton, Tea**
	NATIONAL DAY **October 9**
LANGUAGES **Swahili, English**	ORGANISATIONS **UN, OAU, PTA, C**

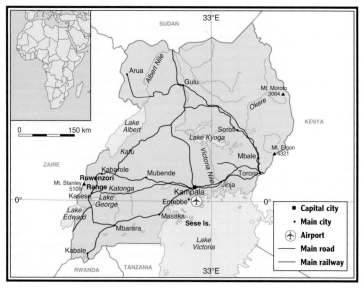

OCCUPIED by UK 1890; independent 1962. Bordered to the west by western arm of African Rift Valley, lakes and mountain ranges; to the east by mountains along Kenyan boundary. Central high plateau is coarse savannah grassland. Warm climate with moderate rainfall. Area around Lake Victoria has been cleared for cultivation. Main crops are coffee, cotton and tea. Great supplies of freshwater fish in Lake Victoria.

UKRAINE

STATUS **Republic**

AREA **603,700 sq km**

CAPITAL **Kiev**

MAIN CITIES **Kharkov, Donetsk, Odessa, Dnepropetrovsk**

POPULATION **51,857,000**
Density **86 people per sq km**
Life Expectancy **71 years**
Infant Mortality **13 per 1000**

LANGUAGES **Ukrainian, Russian**

RELIGIONS **Orthodox**

CURRENCY **Rouble**

MAIN PRODUCTS **Coal, Iron ore, Metal products, Grain**

ORGANISATIONS **UN**

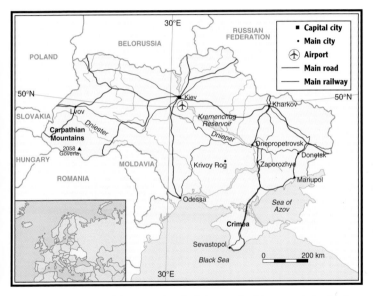

INDEPENDENT 1991. Vast level plains and with mountains in the west and in Crimea. Diverse landscape: marshes, forests, wooded and treeless semi-arid plains. Very fertile soils. Grain, potatoes, vegetables and fruits, industrial crops (notably sugar beet and sunflower seeds) and fodder crops grown. Food processing and viticulture important. Rich reserves of iron ore, coal, lignite, oil and gas. Mining, metal production, machine-building, engineering and chemicals dominate industry. Well-equipped for adaptation to free market conditions.

UNITED ARAB EMIRATES

STATUS **Federation of Emirates**

AREA **75,150 sq km**

CAPITAL **Abu Dhabi**

MAIN CITIES **Dubai, Sharjah, Ras al Khaimah**

POPULATION **1,589,000**
Density **21 people per sq km**
Life Expectancy **71 years**
Infant Mortality **22 per 1000**

LANGUAGES **Arabic**

RELIGIONS **Sunni Muslim**

CURRENCY **Dirham**

GNP **18,430 US $ per person**

MAIN PRODUCTS **Crude oil**

NATIONAL DAY **December 2**

ORGANISATIONS **UN, OPEC, AL**

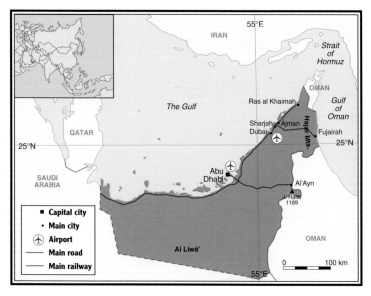

- ■ Capital city
- • Main city
- ✈ Airport
- — Main road
- — Main railway

FORMERLY Trucial States; independent 1971. Seven emirates stretched along the south-eastern shores of The Gulf constitute this oil-rich Arab state. Flat deserts cover most of the landscape; Hajar Mountains in the east. Very high summer temperatures; low winter rainfall. Only desert oases are fertile, producing fruit and vegetables. Trade dominated by exports of oil and natural gas.

UNITED KINGDOM

STATUS **Constitutional Monarchy**

AREA **244,755 sq km**

CAPITAL **London**

MAIN CITIES **Birmingham, Glasgow, Manchester, Cardiff**

POPULATION **57,411,000**
Density **235 people per sq km**
Life Expectancy **76 years**
Infant Mortality **8 per 1000**

LANGUAGES **English**

RELIGIONS **Protestant, Roman Catholic**

CURRENCY **Pound**

GNP **14,570 US $ per person**

MAIN PRODUCTS **Machinery, Transport equipment, Chemical products**

ORGANISATIONS **UN, EC, OECD, NATO, CP, C**

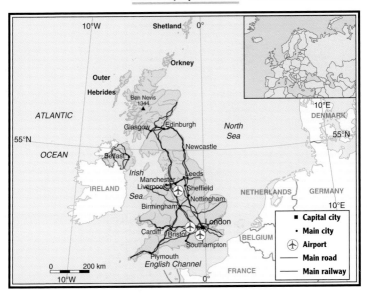

ESTABLISHED 1801 by union of Great Britain and Ireland; Republic of Ireland withdrew 1921. Consist of the island of Britain and part of the island of Ireland, off the north-west coast of Europe. Contrasting landscapes in Britain of mountains to the north and west and lowlands to the south and east. Main agricultural areas and centres of population are in the lowlands. Farm products important though main employment is in service and manufacturing industry. Trade now mainly with EC partners rather than Commonwealth countries.

U.K. Dependencies

GUERNSEY

STATUS	**Crown Dependency**
AREA	**78 sq km**
CAPITAL	**St. Peter Port**
POPULATION	**60,000** Density **769 people per sq km**
LANGUAGES	**English, French**

JERSEY

STATUS	**Crown Dependency**
AREA	**116 sq km**
CAPITAL	**St. Helier**
POPULATION	**84,000** Density **724 people per sq km**
LANGUAGES	**English, French**

GIBRALTAR

STATUS	**Colony**
AREA	**6 sq km**
CAPITAL	**Gibraltar Town**
POPULATION	**31,000** Density **5,167 people per sq km**
LANGUAGES	**English, Spanish**

ISLE OF MAN

STATUS	**Crown Dependency**
AREA	**572 sq km**
CAPITAL	**Douglas**
POPULATION	**64,000** Density **112 people per sq km**
LANGUAGES	**English, Manx**

ANGUILLA

STATUS	**Territory**
AREA	**91 sq km**
CAPITAL	**The Valley**
POPULATION	**8,000**
Density	**88 people per sq km**
LANGUAGES	**English**

BRITISH VIRGIN IS.

STATUS	**Crown Colony**
AREA	**153 sq km**
CAPITAL	**Road Town**
POPULATION	**13,000**
Density	**85 people per sq km**
LANGUAGES	**English**

CAYMAN ISLANDS

STATUS	**Colony**
AREA	**259 sq km**
CAPITAL	**George Town**
POPULATION	**27,000**
Density	**104 people per sq km**
LANGUAGES	**English**

BAHAMAS
70°W
Turks and
Caicos Islands (U.K.)
ATLANTIC OCEAN
CUBA
20°N
20°N
Cayman
Islands (U.K.)
JAMAICA
HAITI
DOMINICAN
REPUBLIC
British
Virgin Is. (U.K.)
Anguilla (U.K.)
Leeward
Islands
Montserrat (U.K.)
Caribbean Sea
Windward
Islands
Lesser Antilles
TRINIDAD
AND
TOBAGO
PANAMA
VENEZUELA
COLOMBIA
70°W
GUYANA

MONTSERRAT

STATUS	**Colony**
AREA	**98 sq km**
CAPITAL	**Plymouth**
POPULATION	**13,000**
Density	**133 people per sq km**
LANGUAGES	**English**

TURKS AND CAICOS IS.

STATUS	**Colony**
AREA	**430 sq km**
CAPITAL	**Cockburn Town**
POPULATION	**10,000**
Density	**23 people per sq km**
LANGUAGES	**English**

UNITED KINGDOM

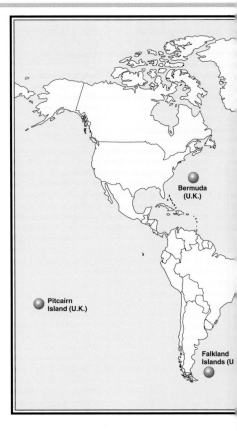

Bermuda
(U.K.)

Pitcairn
Island (U.K.)

Falkland
Islands (U

BERMUDA

STATUS **Colony**

AREA **54 sq km**

CAPITAL **Hamilton**

POPULATION **61,000**
Density **1,130 people per sq km**

LANGUAGES **English**

FALKLAND ISLANDS

STATUS **Colony**

AREA **12,173 sq km**

CAPITAL **Stanley**

POPULATION **2,000**
Density **1 person per sq km**

LANGUAGES **English**

Hong Kong (U.K.)

. Helena
Dependencies
.K.)

HONG KONG

STATUS	**Crown Colony**
AREA	**1,062 sq km**
CAPITAL	**Victoria**
POPULATION **5,801,000** Density **5,462 people per sq km**	
LANGUAGES	**Cantonese,** **English**

PITCAIRN ISLAND

STATUS	**Colony**
AREA	**5 sq km**
CAPITAL	**Adamstown**
POPULATION **50** Density **10 people per sq km**	
LANGUAGES	**English**

ST. HELENA AND DEPENDENCIES

STATUS	**Colony**
AREA	**308 sq km**
CAPITAL	**Jamestown**
POPULATION **7,000** Density **23 people per sq km**	
LANGUAGES	**English**

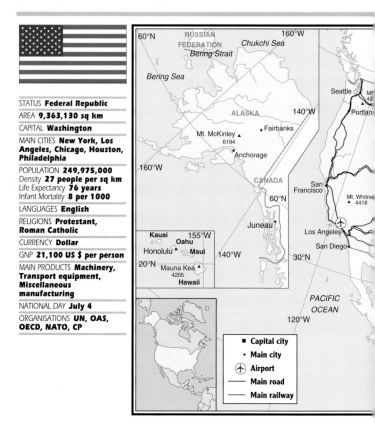

STATUS **Federal Republic**

AREA **9,363,130 sq km**

CAPITAL **Washington**

MAIN CITIES **New York, Los Angeles, Chicago, Houston, Philadelphia**

POPULATION **249,975,000**
Density **27 people per sq km**
Life Expectancy **76 years**
Infant Mortality **8 per 1000**

LANGUAGES **English**

RELIGIONS **Protestant, Roman Catholic**

CURRENCY **Dollar**

GNP **21,100 US $ per person**

MAIN PRODUCTS **Machinery, Transport equipment, Miscellaneous manufacturing**

NATIONAL DAY **July 4**

ORGANISATIONS **UN, OAS, OECD, NATO, CP**

Map labels: 60°N, RUSSIAN FEDERATION, Chukchi Sea, 160°W, Bering Strait, Bering Sea, Seattle, Mt 43, ALASKA, 140°W, Portland, Fairbanks, Mt. McKinley 6194, Anchorage, 160°W, CANADA, San Francisco, Mt. Whitney 4418, 60°N, Juneau, Kauai 155°W, Oahu, Honolulu Maui, 140°W, Los Angeles, San Diego, 20°N, Mauna Kea 4205, Hawaii, 30°N, PACIFIC OCEAN, 120°W

■ Capital city
• Main city
✈ Airport
— Main road
— Main railway

ORIGINAL 13 states independent from UK 1776. World's third largest country with third largest population; very high living standard. Huge variety of terrain, vegetation and climates. Reserves of timber, coal, oil, natural gas, hydro-electric power, copper, lead, zinc, silver and iron ore; also clays, gypsum, lime, phosphate, salt, sand, gravel and sulphur. Advanced industrial nation. Main industries are vehicle manufacture, armaments, machinery, electrical goods, electronics, entertainment, textiles and clothing.

UNITED STATES of AMERICA

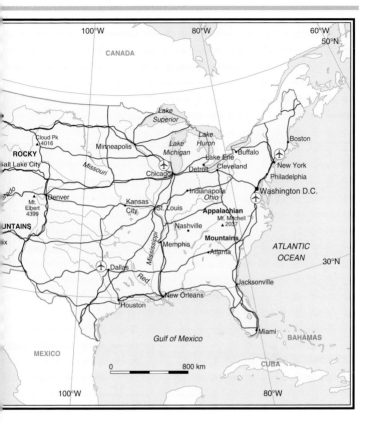

CANADA

100°W   80°W   60°W
50°N

Lake Superior

Cloud Pk ▲4016

ROCKY

Salt Lake City   *Missouri*

Lake Michigan   *Lake Huron*   Buffalo   Boston

Minneapolis   Detroit   Lake Erie   New York

Chicago   Cleveland   Philadelphia

Denver   Mt. Elbert 4399   Kansas City   St. Louis   Indianapolis   *Ohio*   Washington D.C.

Colorado   **Appalachian**

MOUNTAINS   Nashville   Mt. Mitchell ▲2037

Phoenix   Memphis   **Mountains**   ATLANTIC OCEAN   30°N

Mississippi   Atlanta

Dallas   *Red*   Jacksonville

Houston   New Orleans

Gulf of Mexico   Miami   BAHAMAS

MEXICO   0   800 km   CUBA

100°W   80°W

U.S.A. Overseas Territories

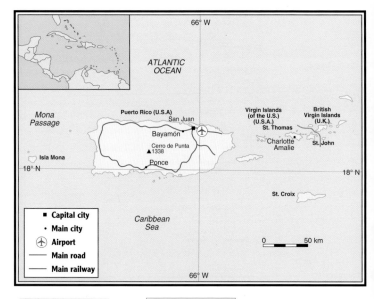

PUERTO RICO

STATUS **Commonwealth**

AREA **8,897 sq km**

CAPITAL **San Juan**

POPULATION **3,599,000**
Density **405 people per sq km**

LANGUAGES **Spanish, English**

VIRGIN IS. (of the U.S.)

STATUS **Unincorporated Territory**

AREA **352 sq km**

CAPITAL **Charlotte Amalie**

POPULATION **117,000**
Density **332 people per sq km**

LANGUAGES **English**

U.S.A. Overseas Territories

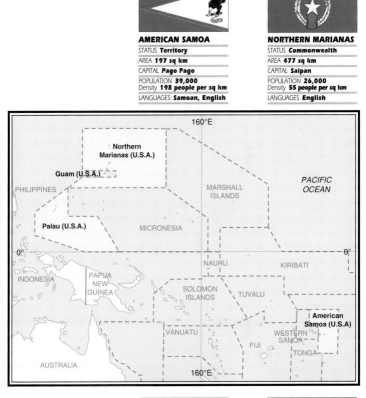

AMERICAN SAMOA
STATUS **Territory**

AREA **197 sq km**

CAPITAL **Pago Pago**

POPULATION **39,000**
Density **198 people per sq km**

LANGUAGES **Samoan, English**

NORTHERN MARIANAS
STATUS **Commonwealth**

AREA **477 sq km**

CAPITAL **Saipan**

POPULATION **26,000**
Density **55 people per sq km**

LANGUAGES **English**

GUAM
STATUS **Unincorporated Territory**

AREA **541 sq km**

CAPITAL **Agana**

POPULATION **119,000**
Density **220 people per sq km**

LANGUAGES **Chamorro, English**

PALAU
STATUS **UN Trust Territory**

AREA **497 sq km**

CAPITAL **Koror**

POPULATION **18,000**
Density **36 people per sq km**

LANGUAGES **English, Palaun**

URUGUAY

STATUS **Republic**	RELIGIONS **Roman Catholic**
AREA **186,925 sq km**	CURRENCY **Peso**
CAPITAL **Montevideo**	GNP **2,620 US $ per person**
MAIN CITIES **Las Piedras**	MAIN PRODUCTS **Meat and meat products, Textiles, Hides and skins**
POPULATION **3,096,000**	
Density **17 people per sq km**	
Life Expectancy **73 years**	NATIONAL DAY **August 25**
Infant Mortality **20 per 1000**	ORGANISATIONS **UN, OAS, ALADI**
LANGUAGES **Spanish**	

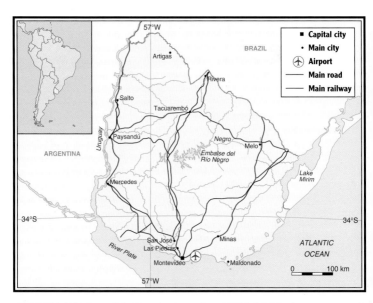

PART OF Spanish South America 1726-1814; incorporated into Brazil as Cisplatine Province 1821; independent 1828. Narrow coastal plain; rolling hills inland. Most land is farmed. Temperate climate and adequate rainfall provide good agricultural potential but most farmland devoted to sheep and cattle grazing. Entire economy relies on production of meat and wool. Principal industry is food processing. Few mineral resources.

UZBEKISTAN

STATUS **Republic**

AREA **447,400 sq km**

CAPITAL **Tashkent**

MAIN CITIES **Samarkand, Andizhan, Namangan**

POPULATION **20,531,000**
Density **46 people per sq km**
Life Expectancy **69 years**
Infant Mortality **38 per 1000**

LANGUAGES **Uzbek**

RELIGIONS **Sunni Muslim**

CURRENCY **Rouble**

MAIN PRODUCTS **Cotton, Rice, Textiles**

ORGANISATIONS **UN**

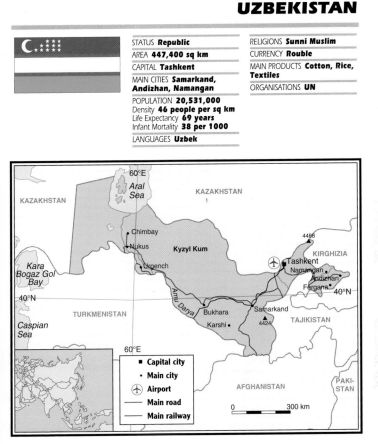

ESTABLISHED 1924 as constituent republic of USSR; independent 1991. Sun-baked lowlands with mountains in the south and east. Continental, very dry climate with abundant sunshine; mild, short winters. Vast reserves of natural gas, oil, coal, iron and other metals. Extensive hydro-electric power. Fertile soils (when irrigated) and good pastures. Well situated for cattle raising and cotton production. Specialises in machinery for cotton cultivation and harvesting, for irrigation projects, road-building and textile processing.

VANUATU

STATUS **Republic**	RELIGIONS **Christian**
AREA **14,765 sq km**	CURRENCY **Vatu**
CAPITAL **Port-Vila**	GNP **820 US $ per person**
POPULATION **147,000** Density **10 people per sq km** Life Expectancy **70 years** Infant Mortality **55 per 1000**	MAIN PRODUCTS **Copra, Meat, Cocoa**
	NATIONAL DAY **July 30**
LANGUAGES **English, French, Bislama**	ORGANISATIONS **UN, SPF, C**

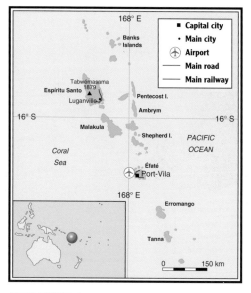

FORMER Anglo-French Condominium of the New Hebrides; independent 1980. Chain of densely forested, mountainous, volcanic islands in South Pacific. Climate is tropical and cyclonic. Copra, cocoa and coffee grown mainly for export. Fish, pigs and sheep important for home consumption as well as yam, taro, manioc and bananas. Manganese is only mineral.

VATICAN CITY

STATUS **Independent Sovereignty**

AREA **0.44 sq km**

POPULATION **1,000**

LANGUAGES **Italian**

RELIGIONS **Roman Catholic**

CURRENCY **Lira**

NATIONAL DAY **October 22**

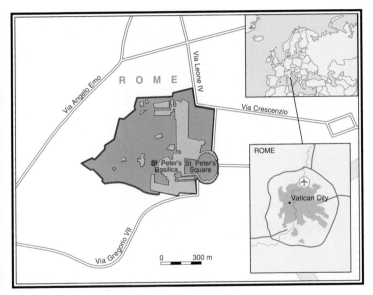

WORLD'S smallest independent state. Headquarters of Roman Catholic Church. Papal residence since 5th century. Destination for pilgrims and tourists from all over the world. Most income derived from voluntary contributions and interest on investments. Only industries are those connected with the Church.

VENEZUELA

STATUS **Federal Republic**	RELIGIONS **Roman Catholic**
AREA **912,045 sq km**	CURRENCY **Bolívar**
CAPITAL **Caracas**	GNP **2,450 US $ per person**
MAIN CITIES **Maracaibo, Valencia, Barquisimeto, Maracay**	MAIN PRODUCTS **Crude oil, Petroleum products, Iron ore**
POPULATION **19,735,000** Density **22 people per sq km** Life Expectancy **70 years** Infant Mortality **33 per 1000**	NATIONAL DAY **July 5** ORGANISATIONS **UN, OAS, OPEC, ALADI**
LANGUAGES **Spanish**	

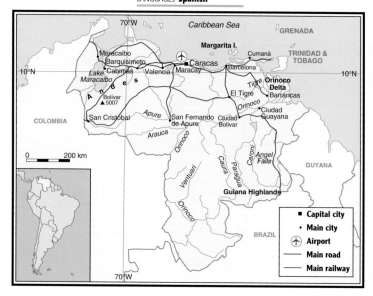

INDEPENDENT from Spain 1821. One of richest countries in Latin America. Andean Mountains in the west; humid lowlands around Lake Maracaibo in the north; coarse-grassed central plains; and extension of Guiana Highlands in the south and east. Tropical climate in the south; warm temperate along northern coasts, where most people live. Economy dominated by oil, which earns most export revenue. Bauxite and iron ore are also important. Most people employed in industrial and manufacturing developments.

VIETNAM

STATUS **People's Socialist Republic**

AREA **329,565 sq km**

CAPITAL **Hanoi**

MAIN CITIES **Ho Chi Minh, Haiphong, Nha Trang, Da Nang**

POPULATION **66,200,000**
Density **201 people per sq km**
Life Expectancy **64 years**
Infant Mortality **54 per 1000**

LANGUAGES **Vietnamese**

RELIGIONS **Buddhist**

CURRENCY **Dong**

GNP **215 US $ per person**

MAIN PRODUCTS **Raw materials, Handicrafts, Agricultural products**

NATIONAL DAY **September 2**

ORGANISATIONS **UN**

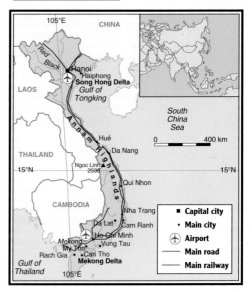

PART OF former French Indo-China; independent 1954. Mountainous backbone and two extensive river deltas: the Song Hong in the north and the Mekong River in the south. Extensive rainforests in central mountainous areas. Monsoons bring moderate rainfall. Main crops are rice (in the north), coffee and rubber. Mineral reserves of coal, lignite, anthracite, iron ore and tin. Industry expanding rapidly, but decades of warfare and internal strife have impeded development.

VIRGIN ISLANDS (U.S.) see page 222
WALLIS AND FUTUNA ISLANDS see page 93

WESTERN SAHARA

STATUS **Republic (de facto controlled by Morocco)**

AREA **266,770 sq km**

CAPITAL **Laâyoune**

POPULATION **179,000**
Density **1 person per sq km**
Life Expectancy **40 years**
Infant Mortality **176 per 1000**

LANGUAGES **Arabic**

RELIGIONS **Sunni Muslim**

CURRENCY **Moroccan Dirham**

MAIN PRODUCTS **Phosphates**

ORGANISATIONS **OAU**

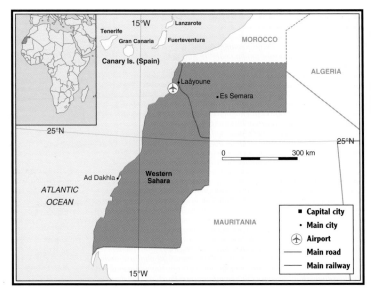

LOCATED in north-west Africa, bordering the Atlantic Ocean. Mainly desert with little agricultural land. Has rich phosphate reserves. Frente Polisario launched armed struggle against Spanish rule in 1973. In 1976 Spain withdrew and the country was partitioned between Morocco and Mauritania. In 1979 Morocco took over all of Western Sahara, but Moroccan sovereignty is not universally recognised. Frente Polisario have continued their struggle for independence. The UN has attempted to oversee a referendum but the situation is at a standstill.

STATUS **Constitutional Monarchy**	RELIGIONS **Protestant**
AREA **2,840 sq km**	CURRENCY **Tala**
CAPITAL **Apia**	GNP **580 US $ per person**
POPULATION **164,000**	MAIN PRODUCTS **Coconut oil, Copra, Taro, Cocoa**
Density **58 people per sq km**	
Life Expectancy **65 years**	NATIONAL DAY **January 1**
Infant Mortality **48 per 1000**	ORGANISATIONS **UN, C**
LANGUAGES **Samoan, English**	

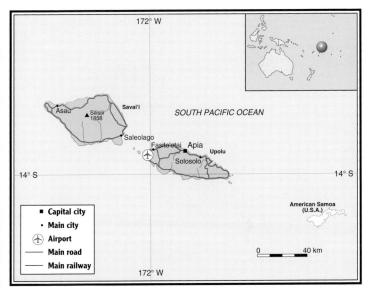

ADMINISTERED by New Zealand since 1920; independent 1962. Comprises nine volcanic, tropical islands in South Pacific Ocean; only four are populated. High rainfall and generally hot. Main exports are copra, coconut oil, taro, cocoa and fruit. Only industries are food processing and timber products. Leading imports are food products, consumer goods, machinery and animals.

YEMEN

STATUS **Republic**	RELIGIONS **Muslim**
AREA **481,155 sq km**	CURRENCY **Rial, Dinar**
CAPITAL **San'a**	GNP **745 US $ per person**
MAIN CITIES **Aden (commercial capital)**	MAIN PRODUCTS **Coffee, Cigarettes, Biscuits, Leather, Grapes**
POPULATION **11,282,000** Density **23 people per sq km** Life Expectancy **53 years** Infant Mortality **107 per 1000**	NATIONAL DAY **May 22**
	ORGANISATIONS **UN, AL**
LANGUAGES **Arabic**	

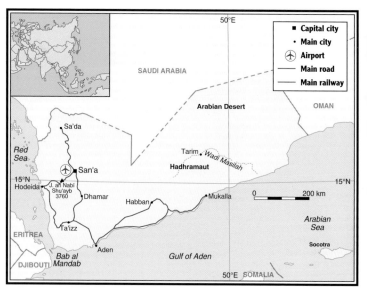

YEMEN Arab Republic and People's Democratic Republic of Yemen unified 1990. North-west is mainly mountainous and relatively wet; here and in the west cereals, cotton, fruits and vegetables are cultivated. Sorghum, millet, wheat and barley grow on southern coastal fringes. In the north and east lies the uninhabited Arabian Desert. Most people are farmers or nomadic herders. Main livestock are sheep, goats, cattle and poultry. Reserves of oil and salt. Industry confined to small-scale manufacturing.

YUGOSLAVIA

STATUS **Federal Republic**

AREA **127,885 sq km**

CAPITAL **Belgrade**

MAIN CITIES **Priština, Subotica, Kragujevac, Podgorica**

POPULATION **10,000,000**
Density **78 people per sq km**
Life Expectancy **73 years**
Infant Mortality **21 per 1000**

LANGUAGES **Serbo-Croat**

RELIGIONS **Orthodox, Roman Catholic, Muslim**

CURRENCY **Dinar**

GNP **2,490 US $ per person**

MAIN PRODUCTS **Coal, Copper, Transport equipment, Chemicals**

NATIONAL DAY **November 29**

ORGANISATIONS **UN**

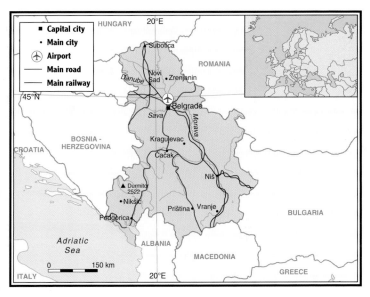

KINGDOM of Serbs, Croats and Slovenes created 1918; federation 1945; Croatia and Slovenia ceded 1991, Bosnia-Herzegovina 1992, Macedonia also effectively a separate country though not universally recognised. Secession of the more advanced northern republics a very severe blow to economy as most industrial installations lie in this area. Conflict between ethnic factions in Bosnia-Herzegovina has wrought further havoc; currency now all but worthless. Economic future of what is left of Yugoslavia looks bleak.

ZAIRE

STATUS **Republic**

AREA **2,345,410 sq km**

CAPITAL **Kinshasa**

MAIN CITIES **Lubumbashi, Kisangani, Kananga**

POPULATION **35,562,000**
Density **15 people per sq km**
Life Expectancy **54 years**
Infant Mortality **75 per 1000**

LANGUAGES **French, Lingala**

RELIGIONS **Roman Catholic**

CURRENCY **Zaire**

GNP **260 US $ per person**

MAIN PRODUCTS **Copper, Coffee, Diamonds, Crude oil, Cobalt**

NATIONAL DAY **November 24**

ORGANISATIONS **UN, OAU**

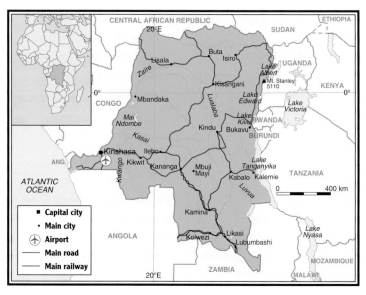

- ■ **Capital city**
- • **Main city**
- ✈ **Airport**
- — **Main road**
- — **Main railway**

BELGIAN colony of Belgian Congo since 1908; independent 1960. Dominated by Zaire River's drainage basin, which is mostly tropical rainforest. Very variable climate is basically equatorial with high temperatures and high rainfall. Soils are poor. Most people practise shifting cultivation. Cassava, cocoa, coffee, cotton, millet, rubber and sugar-cane are grown. Rich in mineral resources including copper, cobalt, diamonds, gold, manganese, uranium and zinc. Abundant wildlife. Tourism becoming important.

STATUS **Republic**	RELIGIONS **Christian**
AREA **752,615 sq km**	CURRENCY **Kwacha**
CAPITAL **Lusaka**	GNP **390 US $ per person**
MAIN CITIES **Kitwe, Ndola, Mufulira, Kabwe**	MAIN PRODUCTS **Copper, Cobalt, Zinc, Tobacco**
POPULATION **7,818,000**	NATIONAL DAY **October 24**
Density **10 people per sq km**	ORGANISATIONS **UN, OAU, PTA, SADC, C**
Life Expectancy **55 years**	
Infant Mortality **72 per 1000**	
LANGUAGES **English**	

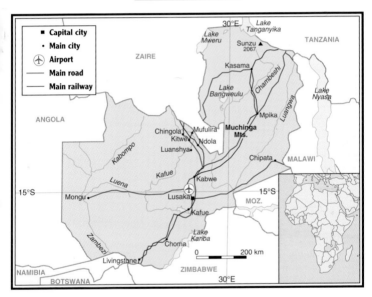

NORTHERN Rhodesia UK colony since 1889; Federation of Rhodesia and Nyasaland 1953-63; independent 1964. Lies in interior of south-central Africa. Mainly high rolling plateaux. Altitude moderates potentially tropical climate. North receives moderate rainfall; the south, less. Most of Zambia is grassland, with some forest in the north. Subsistence farming dominates. Copper mainstay of economy, though reserves are fast running out. Lead, zinc, cobalt and tobacco also important. Diverse, abundant wildlife encourages tourism.

ZIMBABWE

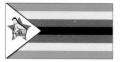

STATUS **Republic**	RELIGIONS **Christian, Traditional**
AREA **390,310 sq km**	CURRENCY **Dollar**
CAPITAL **Harare**	GNP **640 US $ per person**
MAIN CITIES **Bulawayo, Chitungwiza, Gweru**	MAIN PRODUCTS **Tobacco, Gold, Nickel**
POPULATION **9,369,000**	NATIONAL DAY **April 18**
Density **24 people per sq km**	ORGANISATIONS **UN, OAU, PTA, SADC, C**
Life Expectancy **61 years**	
Infant Mortality **55 per 1000**	
LANGUAGES **English**	

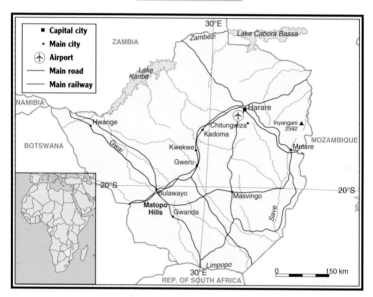

UK **COLONY** of Southern Rhodesia since 1923; Federation of Rhodesia and Nyasaland 1953-63; renamed Rhodesia; unilateral declaration of independence 1965; independent 1980. Rolling central plateaux with Zambezi and Limpopo river valleys in north and south respectively. Altitude moderates tropical climate. Reserves of chrome, nickel, platinum and coal; gold and asbestos especially important. Maize is main crop; also staple food for most people. Tobacco, tea, sugar-cane and fruit also grown. Manufacturing industry slowly developing.